£3.20

O/s 26 90101

5/75

Principles of
Transistor
Circuits

Principles of Transistor Circuits

INTRODUCTION TO THE DESIGN OF
AMPLIFIERS, RECEIVERS AND DIGITAL CIRCUITS

S. W. AMOS, B.Sc., C.Eng., M.I.E.E.

formerly Head of Technical Publications Section,
Engineering Training Department,
British Broadcasting Corporation

LONDON
NEWNES–BUTTERWORTHS

THE BUTTERWORTH GROUP

ENGLAND
Butterworth & Co (Publishers) Ltd
London: 88 Kingsway, WC2B 6AB

AUSTRALIA
Butterworths Pty Ltd
Sydney: 586 Pacific Highway, NSW 2067
Melbourne: 343 Little Collins Street, 3000
Brisbane: 240 Queen Street, 4000

CANADA
Butterworth & Co (Canada) Ltd
Scarborough, Ontario: 2265 Midland Avenue, M1P 4S1

NEW ZEALAND
Butterworths of New Zealand Ltd
Wellington: 26–28 Waring Taylor Street, 1

SOUTH AFRICA
Butterworth & Co (South Africa) (Pty) Ltd
Durban: 152–154 Gale Street

First published for *Wireless World* by Iliffe Books Ltd, 1959
Second edition 1961
Third edition 1965
Fourth edition 1969
Second impression 1972
Fifth edition 1975 by Newnes–Butterworths
an imprint of the Butterworth Group
© S. W. Amos, 1975

ISBN 0 408 00160 7 Standard
     0 408 00161 5 Limp

Filmset and printed Offset Litho in England by
Cox & Wyman Ltd, London, Fakenham and Reading

Preface

Although no dramatic changes in transistor circuitry have occurred since the fourth edition of this book was published in 1969, there has been an increase in the use of field-effect transistors as discrete components and in integrated circuits. There has also been more widespread use of integrated circuits in sound and television receivers and in digital equipment.

To keep the book up-to-date this new edition has more information on circuits using f.e.t.s and the treatment of switching circuits has been expanded to cover the principles of digital equipment in general. A number of minor changes have been made to eliminate dated information or to simplify or clarify the text.

S.W.A.
Broadway, Worcs.

Acknowledgments

The author is grateful to his former employers, the British Broadcasting Corporation for permission to use the circuit diagrams of two BBC-designed amplifiers. Thanks are also due to Thorn-A.E.I. Radio Valves and Tubes Ltd. who provided valuable information on the design of i.f. amplifiers (particularly those using silicon transistors) and to Mullard Limited for information on bipolar and field-effect transistors, the design of video and colour-difference amplifiers and on electric-motor control by thyristors. The information on monolithic amplifiers is used by kind permission of Motorola Semiconductor Products Inc. and on the u.h.f. tuner by permission of Rank-Bush-Murphy Ltd.

The following organisations also provided useful information:

General Electric Co. Ltd.
Ferguson Radio Corporation Ltd.
Philips Electrical Ltd.
Texas Instruments Ltd.
S.G.S. Fairchild Limited.

The author also gratefully acknowledges the help of Dr. J. R. Tillman who made useful comments on Appendix A.

Contents

CHAPTER 1

Semiconductors and Junction Diodes

INTRODUCTION

The most significant development in electronics since the Second World War has been the introduction of the transistor. This minute semiconductor device can amplify, oscillate and be used for switching and other purposes. In all its applications it has an efficiency much greater than that of a thermionic valve. Moreover the transistor has a longer life than a valve, is non-microphonic and is much cheaper than a valve.

It is not surprising that transistors have superseded valves in most categories of electronic equipment. They will probably supplant valves completely in all except high-power applications.

Early germanium transistors had limited output power and operated satisfactorily only up to a few MHz: moreover they had appreciable leakage currents which caused difficulties at high ambient temperatures. Later developments particularly the introduction of the silicon epitaxial planar transistor eliminated these limitations. Modern transistors can give more than 25 W output; they operate satisfactorily up to frequencies above 1,000 MHz, and leakage currents are no longer troublesome. The characteristics are however dependent on temperature and precautions are necessary to ensure stability of operating conditions and gain.

By its physical structure the bipolar transistor is a current-controlled device, that is to say its output current is linearly related to the input current. The input resistance is low (and dependent on signal amplitude) and the output resistance is high. The associated circuits must be designed to operate with such resistances

1

and thus differ from those used with valves. On the other hand the properties of the field-effect transistor are very similar to those of the valve. It is the purpose of this book to give the fundamental principles of the design of circuits for both types of transistor.

To explain the origin of the properties of transistors it is useful to begin with an account of the physics of semiconductors.

DEFINITION OF A SEMICONDUCTOR

The heart of a transistor consists of semiconducting material, e.g. germanium or silicon and the behaviour of the transistor largely depends on the properties of this material. As the name suggests a semiconducting material is one with a conductivity lying between that of an insulator and that of a conductor: that is to say one for which the resistivity lies between, say 10^{12} Ω-cm (a value typical of glass) and 10^{-6} Ω-cm (approximately the value for copper). Typical values for the resistivity of a semiconducting material lie between 1 and 100 Ω-cm.

Such a value of resistivity could, of course, be obtained by mixing a conductor and an insulator in suitable proportions but the resulting material would not be a semiconductor. Another essential feature of a semiconducting material is that its electrical resistance

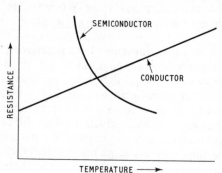

Fig. 1.1. *Resistance-temperature relationship for a conductor and a semiconductor*

decreases with increase in temperature over a particular temperature range which is characteristic of the semiconductor. This behaviour contrasts with that of elemental metallic conductors for which the resistance increases with rise in temperature. This is illustrated in Fig. 1.1, which gives curves for a conductor and a semiconductor. The curve for the conductor shows the resistance increasing linearly with increase in temperature, whereas that for the semiconductor shows the resistance decreasing exponentially with increase in temperature. Over the significant temperature range the relationship between resistance and temperature for a semiconductor could be written

$$R_t = ae^{b/T}$$

where R_t is the resistance at an absolute temperature T, a and b being constants characteristic of the semiconducting material. The two curves in Fig. 1.1 are not to the same vertical scale of resistance.

All semiconducting materials exhibit the temperature dependence discussed in the paragraphs above in the pure state: the addition of impurities raises the temperature at which the material exhibits this behaviour, i.e. the region of negative temperature coefficient.

Germanium in its pure state is a poor conductor, the resistivity being 46 Ω-cm at 27°C, and is of little direct use in transistor manufacture. However, by the addition of a very small but definite amount of a particular type of impurity, the resistivity can be reduced and the material made suitable for transistors. Germanium and silicon so treated are extensively employed in the manufacture of transistors.

The behaviour of semiconductors can be explained in terms of atomic theory. The atom is assumed to have a central nucleus which carries most of the mass of the atom and has a positive charge. A number of electrons carrying a negative charge revolve around the nucleus. The total number of electrons revolving around a

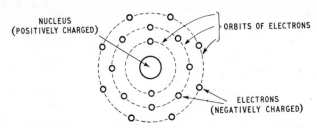

Fig. 1.2. Simplified diagram of structure of atom: for simplicity, electron orbits are shown as circular and co-planar

particular nucleus is sufficient to offset the positive nuclear charge, leaving the atom electrically neutral. The number of electrons associated with a given nucleus is equal to the atomic number of the element. The electrons revolve in a number of orbits and, for the purpose of this discussion, the orbits may be regarded as concentric, the nucleus being at the centre, as shown in Fig. 1.2.

This diagram is greatly simplified; the orbits are in practice neither concentric nor co-planar.

The first orbit (sometimes called a ring or a shell) is complete when it contains 2 electrons, and an atom with a single complete shell is that of the inert gas, helium. The second ring is complete when it has 8 electrons, and the atom with the first 2 rings complete is that of the inert gas, neon. The third ring is stable when it has 8 or 18 electrons, and the atom having 2, 8 and 8 electrons in the 1st, 2nd and 3rd rings is that of the inert gas, argon. All the inert gases have their outermost shells stable. It is difficult to remove any electrons from a stable ring or to insert others into it. Atoms combine by virtue of the electrons in the outermost rings: for example an atom with one electron in the outermost ring will willingly combine with another whose outermost ring requires one electron for completion.

The inert gases, having their outer shells stable, cannot combine with other atoms or with each other. The number of electrons in the outermost ring or the number of electrons required to make the outermost ring complete has a bearing on the chemical valency of the element and the outermost ring is often called the *valence ring*.

Now consider the copper atom: it has 4 rings of electrons, the

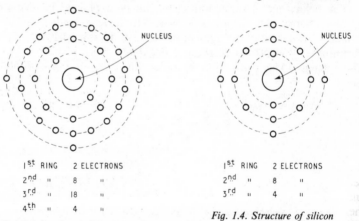

1st RING	2 ELECTRONS		1st RING	2 ELECTRONS
2nd "	8 "		2nd "	8 "
3rd "	18 "		3rd "	4 "
4th "	4 "			

Fig. 1.3 Structure of germanium atom

Fig. 1.4. Structure of silicon atom

first 3 being complete and the 4th containing 1 electron, compared with the 32 needed for completion. Similarly the silver atom has 5 rings, 4 stable and the 5th also containing 1 out of 50 needed for completion. The atoms of both elements thus contain a single electron and this is loosely bound to the nucleus. It can be removed with little effort and is termed a *free electron*. A small e.m.f. applied to a collection of these atoms can set up a stream of free electrons, i.e. an electric current through the metal. Elements in which such free electrons are available are good electrical conductors.

It might be thought that an atom with 17 electrons in the outermost orbit would be an even better conductor, but this is not so. If one electron is added to such an orbit it becomes complete and a great effort is needed to remove it again.

The arrangement of orbital electrons in a germanium atom is pictured in Fig. 1.3. There are 4 rings, the first containing 2 electrons, the second 8, the third 18, and the fourth (final) 4. The total number of electrons is 32, the atomic number of germanium. The corresponding diagram for the silicon atom is given in Fig. 1.4, the three rings containing 2, 8 and 4 electrons respectively. The total number of electrons per atom is 14, the atomic number for silicon. A significant feature of these two atomic structures is that the outermost ring contains 4 electrons: both elements belong to Group IV of the Periodic Table.

Covalent Bonds

It might be thought that some of the 4 electrons in the valence ring of the germanium or silicon atom could easily be displaced and that these elements would therefore be good conductors. In fact, crystals of pure germanium and pure silicon are very poor conductors. To understand this we must consider the relationships between the valence electrons of neighbouring atoms when these are arranged in a regular geometric pattern as in a crystal. The valence electrons of each atom form bonds, termed *covalent bonds*, with those of neighbouring atoms as suggested in Fig. 1.5. It is difficult to portray a three-dimensional phenomenon in a two-dimensional diagram, but the diagram does show the valence electrons oscillating between two neighbouring atoms. The atoms behave in some respects as though each outer ring had 8 electrons and was stable. There are no free electrons and such a crystal is therefore an insulator: this is true of pure germanium and pure silicon at a very low temperature.

At room temperatures, however, germanium and silicon crystals do have a small conductivity even when they are as pure as modern chemical methods can make them. This is partly due to the presence of minute traces of impurities (the way in which these increase conductivity is explained below) and partly because thermal agitation enables some valence electrons to escape from their covalent bonds and thus become available as charge carriers. They are able to do this by virtue of their kinetic energy which, at normal temperatures, is sufficient to allow a very small number to break these bonds. If their kinetic energy is increased by the addition of light or by increase in temperature, more valence electrons escape and the conductivity increases. When the temperature of germanium is raised to 100°C, the conductivity is so great that it swamps

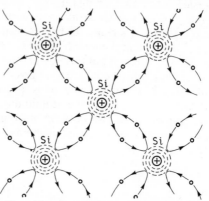

Fig. 1.5. Illustrating covalent bonds in a crystal of pure silicon: for simplicity only electrons in the valence rings are shown

normal transistor action. Moreover, if a reasonable life is required, it is recommended that germanium transistors should not be operated above say 80°C. The life of a germanium transistor is shortened if it is operated above this temperature but a silicon transistor will give a satisfactory life even when operated at 150°C.

Donor Impurities

Suppose an atom of a Group-V element such as arsenic is introduced into a crystal of pure silicon. The atom enters into the lattice structure, taking the place of a silicon atom. Now the arsenic atom has 5 electrons in its outermost orbit and 4 of these form covalent bonds with the electrons of neighbouring atoms as shown in Fig. 1.6. The remaining (5th) electron is left unattached; it is a free electron which can be made to move through the crystal by an e.m.f., leaving a positively-charged ion. These added electrons give the crystal much better conductivity than pure silicon and the added element is termed a *donor* because it gives free electrons to the crystal. Silicon so treated with a Group-V element is termed n-type because negatively-charged particles are available to carry charge through the crystal. It is significant that the addition of the arsenic or some other Group-V element was necessary to give this improvement in conductivity. The added element is often called an impurity and in the language of the chemist it undoubtedly is. However, the word is unfortunate in this context because it suggests that the pentavalent element is unwanted; in fact, it is essential.

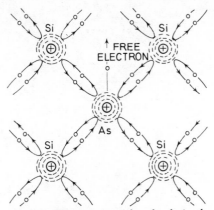

Fig. 1.6. *Illustrating covalent bonds in the neighbourhood of an atom of a Group-V element introduced into a crystal of pure silicon. For simplicity only electrons in the valence rings are shown*

When a battery is connected across a crystal of n-type semiconductor the free electrons are attracted towards the battery positive terminal and repelled from the negative terminal. These forces cause a drift of electrons through the crystal from the negative to the positive terminal: for every electron leaving the crystal to enter the positive terminal another must be liberated from the negative terminal to enter the crystal. The stream of electrons through the crystal constitutes an electric current. If the voltage applied to the crystal is varied the current varies also in direct proportion, and if the battery connections are reversed the direction of the current through the crystal also reverses but it does not change in amplitude; that is to say the crystal is a *linear* conductor.

Acceptor Impurities

Now suppose an atom of a Group-III element such as boron is introduced into a crystal of pure silicon. It enters the lattice structure, taking the place of a silicon atom, and the 3 electrons in the valence ring of the boron atom form covalent bonds with the valence electrons of the neighbouring silicon atoms. To make up the number of covalent bonds to 4, each boron atom competes with a neighbouring atom and may leave this deficient of one electron as shown in Fig. 1.7. A group of covalent bonds, which is deficient

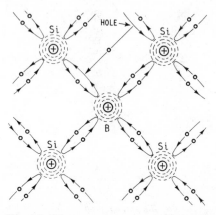

Fig. 1.7. Illustrating covalent bonds in the neighbourhood of an atom of a Group-III element, introduced into a crystal of pure silicon. For simplicity, only electrons in the valence rings are shown

of one electron, behaves in much the same way as a positively-charged particle with a charge equal in magnitude to that of an electron. Such a particle is called a *hole* in semi-conductor theory, and we may say that the introduction of the Group-III impurity gives rise to holes in a crystal of pure silicon. These can carry charge through the crystal and, because these charge-carriers have a positive sign, silicon treated with a Group-III impurity is termed p-type. Such an impurity is termed an *acceptor* impurity because it takes electrons from the silicon atoms. Thus the introduction of the Group-III element into a crystal lattice of pure silicon also increases the conductivity considerably and, when a battery is connected across a crystal of p-type silicon, a current can flow through it in the following manner.

The holes have an effective positive charge, and are therefore attracted towards the negative terminal of the battery and repulsed by the positive terminal. They therefore drift through the crystal from the positive to the negative terminal. Each time a hole reaches the negative terminal, an electron is emitted from this terminal into the hole in the crystal to neutralise it. At the same time an electron from a covalent bond enters the positive terminal to leave another hole in the crystal. This immediately moves towards the negative terminal, and thus a stream of holes flows through the crystal from the positive to the negative terminal. The battery thus loses a steady stream of electrons from the negative terminal and receives a similar stream at its positive terminal. It may be said that a stream of electrons has passed through the crystal from the negative to the positive terminal. A flow of holes is thus equivalent to a flow of electrons in the opposite direction.

If the battery voltage is varied the current also varies in direct proportion: thus p-type silicon is also a linear conductor.

It is astonishing how small the impurity concentration must be to make silicon suitable for use in transistors. A concentration of 1 part in 10^6 may be too large, and concentrations commonly used are of a few parts in 10^8. A concentration of 1 part in 10^8 increases the conductivity by 16 times. Before such a concentration can be introduced, the silicon must first be purified to such an extent that any impurities still remaining represent concentrations very much less than this. Purification is one of the most difficult processes in the manufacture of transistors.

Intrinsic and Extrinsic Semiconductor

If a semiconductor crystal contains no impurities, the only charge carriers present are those produced by thermal breakdown of the covalent bonds. The conducting properties are thus characteristic of the pure semiconductor. Such a crystal is termed an *intrinsic* semiconductor.

In general, however, the semiconductor crystals contain some Group-III and some Group-V impurities, i.e. some donors and some acceptors are present. Some free electrons fit into some holes and neutralise them but there are some residual charge carriers left. If these are mainly electrons they are termed *majority carriers* (the holes being *minority carriers*), and the material is n-type. If the residual charge carriers are mainly holes, these are majority carriers (the electrons being minority carriers) and the semiconductor is termed p-type. In an n-type or p-type crystal the impurities are chiefly responsible for the conduction, and the material is termed an *extrinsic* semiconductor.

PN JUNCTIONS

As already mentioned, an n-type or p-type semiconductor is a linear conductor, but if a crystal of semiconductor has n-type conductivity at one end and p-type at the other end, as indicated in Fig. 1.8, the crystal so produced has asymmetrical conducting properties. That is to say, the current which flows in the crystal when an e.m.f. is applied between the ends depends on the polarity of the e.m.f., being small when the e.m.f. is in one direction and large when it is reversed. Crystals with such conductive properties have obvious applications as detectors or rectifiers.

It is not possible, however, to produce a structure of this type by placing a crystal of n-type semiconductor in contact with a crystal

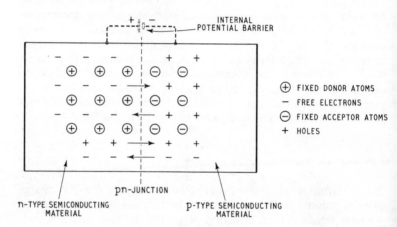

Fig. 1.8. *Pattern of fixed and mobile charges in the region of a pn junction*

of p-type semiconductor. No matter how well the surfaces to be placed together are planed, or how perfect the contact between the two appears, the asymmetrical conductive properties are not properly obtained. The usual way of achieving a structure of this type is by treating one end of a single crystal of n-type semiconductor with a Group-III impurity so as to offset the n-type conductivity at this end and to produce p-type conductivity instead at this point. Alternatively, of course, one end of a p-type crystal could be treated with a Group-V impurity to give n-type conductivity at this end. The semiconducting device so obtained is termed a junction diode, and the non-linear conducting properties can be explained in the following way.

Behaviour of a pn Junction

Fig 1.8 represents the pattern of charges in a crystal containing a pn junction. The ringed signs represent charges due to the impurity atoms and are fixed in position in the crystal lattice: the unringed signs represent the charges of the free electrons and holes (majority carriers) which are liberated by the impurities. The n-region also contains a few holes and the p-region also a few free-electrons: these are minority carriers which are liberated by thermal dissociation of the covalent bonds of the semiconducting element itself.

Even when no external connections are made to the crystal, there is a tendency, due to diffusion, for the free electrons of the n-region to cross the junction into the p-region: similarly the holes in the p-region tend to diffuse into the n-region. However the moment any of these majority carriers cross the junction, the electrical neutrality of the two regions is upset: the n-region loses electrons and gains holes, both causing it to become positively charged with respect to the p-region. Thus a potential difference is established across the junction and this discourages further majority carriers from crossing the junction: indeed only the few majority carriers with sufficient energy succeed in crossing. The potential difference is, however, in the right direction to encourage minority carriers to cross the junction and these cross readily in just sufficient numbers to balance the subsequent small flow of majority carriers. Thus the balance of charge is preserved even though the crystal has a potential barrier across the junction. In Fig. 1.8 the internal potential barrier is represented as an external battery and is shown in dotted lines.

The potential barrier tends to establish a carrier-free zone, known as a *depletion area*, at the junction. The depletion area is similar to the dielectric in a charged capacitor.

Reverse-bias Conditions

Suppose now an external battery is connected across the junction, the negative terminal being connected to the p-region and the positive terminal to the n-region as shown in Fig. 1.9. This con-nection gives a *reverse-biased* junction. The external battery is in parallel with and aiding the fictitious battery, increasing the potential barrier across the junction and the width of the depletion area. Even the majority carriers with the greatest energy now find it almost impossible to cross the junction. On the other hand the minority carriers can cross the junction as easily as before and a steady stream of these flows across. When the minority carriers cross the junction they are attracted by the battery terminals and can then flow as a normal electric current in a conductor. Thus a

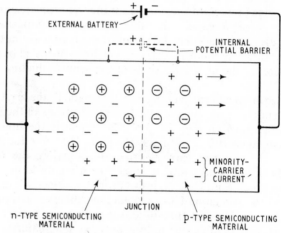

Fig. 1.9. Reverse-bias conditions in a pn junction

current, carried by the minority carriers and known as the *reverse current*, flows across the junction. It is a small current because the number of minority carriers is small: it increases as the battery voltage is increased as shown in Fig. 1.10 but at a reverse voltage

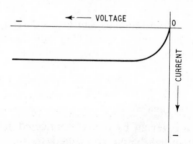

Fig. 1.10. Current-voltage relationship for a reverse-biased pn junction

of less than 1 V becomes constant: this is the voltage at which the rate of flow of minority carriers becomes equal to the rate of production of carriers by thermal breakdown of covalent bonds. Increase in the temperature of the crystal produces more minority carriers and an increase in reverse current. A significant feature of the reverse-biased junction is that the width of the depletion area is controlled by the reverse bias, increasing as the bias increases.

Forward-bias Conditions

If the external battery is connected as shown in Fig. 1.11, with the positive terminal connected to the p-region and the negative terminal to the n-region, the junction is said to be *forward-biased*. The external battery now opposes and reduces the potential barrier

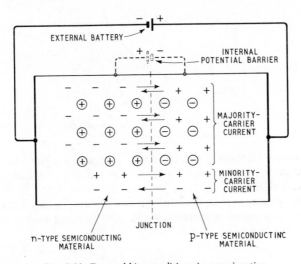

Fig. 1.11. *Forward-bias conditions in a pn junction*

due to the fictitious battery and the majority carriers are now able to cross the junction more readily. A steady flow of majority electrons and majority holes can now flow across the junction and these together constitute a considerable current from the external battery. The depletion area has now disappeared.

The flow of minority carriers across the junction also continues as in reverse-bias conditions but at a reduced scale and these give rise to a second current also taken from the battery but in the opposite direction to that carried by the majority carriers. Except for very small external battery voltages, however, the minority-carrier current is very small compared with the majority-carrier current and can normally be neglected in comparison with it.

The relationship between current and forward-bias voltage is illustrated in Fig. 1.12. The curve has a small slope for small voltages because the internal potential barrier discourages movements of majority carriers across the junction. Increase in applied voltage tends to offset the internal barrier and current increases at a greater

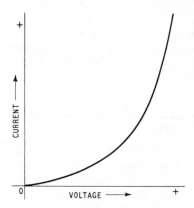

Fig. 1.12. Current-voltage relationship for a forward-biased pn junction

rate. Further increase in voltage almost completely offsets the barrier and gives a steeply-rising current. The curve is, in fact, closely exponential in form.

A pn junction thus has asymmetrical conducting properties, allowing current to pass freely in one direction but hardly at all in the reverse direction.

JUNCTION DIODES

Junction diodes are very efficient and therefore little heat is generated in them in operation. Such diodes can therefore rectify surprisingly large currents: for example a silicon diode with a junction area less than $\frac{1}{2}$ in. in diameter can supply 50 A at 100 V. The capacitance between the terminals of a small-area junction is low enough for it to make an efficient detector.

Avalanche Effect

When a pn junction is reverse-biased the current is carried solely by the minority carriers, and at a given temperature the number of minority carriers is fixed. Ideally, therefore, we would expect the reverse current for a pn junction to rise to a saturation value as the voltage is increased from zero and then to remain constant and independent of voltage, as shown in Fig. 1.10. In practice, when the reverse voltage reaches a particular value which can be 100 V or more the reverse current increases very sharply as shown in Fig. 1.13, an effect known as breakdown. The effect is reproducible, breakdown in a particular junction always occurring at the same value of reverse voltage. This is known as the *Avalanche effect* and reversed-biased diodes known as Avalanche diodes (sometimes called—perhaps incorrectly—Zener diodes) can be used as the basis of a voltage stabiliser circuit. The junction diodes used for this purpose are usually silicon types and examples of voltage stabilising circuits employing such diodes are given in Chapter 16.

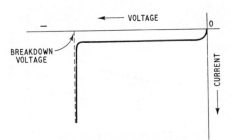

Fig. 1.13. Breakdown in a reverse-biased pn junction

The explanation of the Avalanche effect is thought to be as follows. The reverse voltage applied to a junction diode establishes an electric field across the junction and minority electrons entering it from the p-region are accelerated to the n-region as illustrated in Fig. 1.9. When this field exceeds a certain value some of these electrons collide with valence electrons of the atoms fixed in the crystal lattice and liberate them, thus creating further hole-electron pairs. Some new carriers are themselves accelerated by the electric field due to the reverse bias and in turn collide with other atoms, liberating still further holes and electrons. In this way the number of charge carriers increases very rapidly: the process is, in fact, regenerative. This multiplication in the number of charge carriers produces the sharp increase in reverse current shown in Fig. 1.13. Once the breakdown voltage is exceeded, a very large reverse current can flow and unless precautions are taken to limit this current the junction can be damaged by the heat generated in it. Voltage stabilising circuits using Avalanche diodes must therefore include protective measures to avoid damage due to this cause.

Capacitance Diode (*Varactor Diode*)

As pointed out above, the application of reverse bias to a pn junction discourages majority carriers from crossing the junction, and produces a depletion area, the width of which can be controlled by the magnitude of the reverse bias. Such a structure is similar to that of a charged capacitor and, in fact, a reverse-biased junction diode has the nature of a capacitance shunted by a high resistance. The value

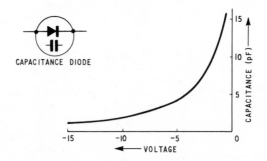

Fig. 1.14. Symbol and typical capacitance-voltage characteristic for a capacitance diode

of the capacitance is dependent on the reverse-bias voltage and can be varied over wide limits by alteration in the bias voltage. This is illustrated in the curve of Fig. 1.14: the capacitance varies with the voltage according to a law of the type

$$C \approx kV^{-1/n}$$

where k is a constant and n is between 2 and 3. When n is 2 the capacitance is inversely proportional to the square root of the voltage. A voltage-sensitive capacitance such as this has a number of useful applications: it can be used as a frequency modulator, as a means of remote tuning in receivers or for a.f.c. purposes in receivers. An example of one of these applications of the reverse-biased junction diode is given in Chapter 16.

Zener Effect

Some reverse-biased junction diodes exhibit breakdown at a very low voltage, say below 5 V. In such examples breakdown is thought to be due, not to Avalanche effect, but to Zener effect which does not involve ionisation by collision. Zener breakdown is attributed to spontaneous generation of hole-electron pairs within the junction region from the inner electron shells. Normally this region is carrier-free but the intense field established across the region by the reverse bias can produce carriers which are then accelerated away from the junction by the field, so producing a reverse current.

Voltage Reference Diode

The breakdown voltage of a reverse-biased junction diode can be placed within the range of a few volts to several hundred volts but for stabiliser and voltage reference applications it is unusual to employ a diode with a breakdown voltage exceeding a few tens of volts. Some of the reasons for this are given below.

The breakdown voltage varies with temperature, the coefficient of variation being negative for diodes with breakdown voltages less than approximately 5·3 V and positive for diodes with breakdown voltages exceeding approximately 6·0 V. Diodes with breakdown voltages between these two limits have very small coefficients of variation and are thus well suited for use in voltage stabilisers. However, for voltage reference purposes, the slope resistance of the breakdown characteristic must be very small and the slope resistance is less for diodes with breakdown voltages exceeding 6 V than for those with lower breakdown voltages. Where variations in temperature are likely to occur it is probably best to use a diode with a breakdown voltage between 5·3 and 6·0 V for voltage reference purposes but if means are available for stabilising the temperature it is probably better to use a diode with a higher breakdown voltage to obtain a lower slope resistance. Diodes with breakdown voltages around 6·8 V have a temperature coefficient (2·5 mV/°C) which matches that of transistors; this can be useful in designing stabilised power supplies.

Voltage reference diodes are usually marketed with preferred values of breakdown voltage (4·7, 5·6, 6·8 V, etc.) and with tolerances of 5 per cent or 10 per cent.

Early voltage reference diodes were rated for 50 mW dissipation but more modern types withstand 300 W. Large diodes are used to protect radars, communications systems and delicate instruments from large electrical transients from nearby electrical equipment.

BACKWARD DIODE

If the breakdown voltage is made very low, the region of low slope resistance virtually begins at the origin. Such a junction has a reverse resistance lower than the forward resistance and can be used as a diode which, by contrast with normal diodes, has low resistance when the p-region is biased negatively relative to the n-region. Such

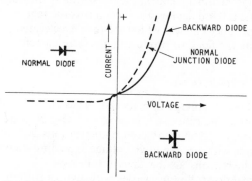

Fig. 1.15. Characteristic curves for a backward diode and normal junction diode. Inset shows graphical symbols used in circuit diagrams

backward diodes, manufactured with low capacitance, are used as microwave detectors. The current-voltage characteristic for a backward diode is shown in Fig. 1.15 and for comparison the curve for a normal junction diode is also included.

TUNNEL DIODE

A pn junction with a very thin junction region, i.e. a very sharp transition from n-type to p-type material, has a characteristic shaped as shown in the solid line in Fig. 1.16: for comparison the characteristic for a diode with a thicker junction region is given in dashed lines. As would be expected breakdown for the thin-junction diode occurs at a very low value of reverse bias and indeed there is in effect no region of high reverse resistance. An unexpected feature of the characteristic is, however, the region of negative slope which occurs at a forward bias: this was first reported by Esaki in 1958. Such a pn junction can be manufactured with very low capacitance and oscillators making use of the negative-resistance kink can function at frequencies as high as thousands of MHz. These diodes are termed tunnel diodes and offer great promise as very high-frequency oscil-

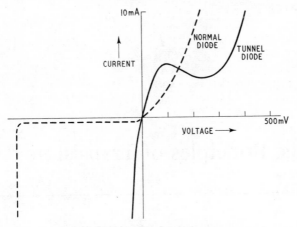

Fig. 1.16. Characteristic of a tunnel diode (solid) compared with that of a normal junction diode (dashed)

lators although the power output is very limited. They can also be used as high-frequency amplifiers but because the tunnel diode has only two terminals there is no isolation between the input and output circuits and it is very difficult to construct a cascaded amplifier.

LIGHT-EMITTING DIODES (LEDS)

When a pn junction is forward-biased electrons are driven into the p-region and holes into the n-region as shown in Fig. 1.11. Some of these charge carriers combine in the junction area and in some of the combinations energy is given out in the form of light. By using an alloy of gallium, arsenic and phosphorus as the semiconducting material the emitted light can be made any colour between red and green but maximum electrical-optical efficiency is obtained when the light is red. Lamps and digital displays using this principle are known as light-emitting diodes. Such lamps typically consume 25 mA at 4 V and have a very long life.

CHAPTER 2

Basic Principles of Transistors

Introduction

Chapter 1 showed that a junction between n-type and p-type materials has asymmetrical conducting properties enabling it to be used for rectification. A bipolar transistor includes two such junctions arranged as shown in Fig. 2.1. Fig. 2.1(a) illustrates one basic type consisting of a layer of n-type material sandwiched between two

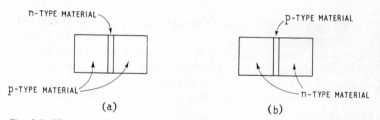

Fig. 2.1. Theoretical diagrams illustrating the structure of (a) a pnp and (b) an npn transistor

layers of p-type material: such a transistor is referred to as a pnp type.

A second type, illustrated in Fig. 2.1(b) has a layer of p-type material sandwiched between two layers of n-type semiconducting material: such a transistor is referred to as an npn type.

In both types, for successful operation, the central layer must be thin. However, it is not possible to construct bipolar transistors by placing suitably-treated layers of semiconducting material in contact. One method which is employed is to start with a single crystal of, say, n-type germanium and to treat it so as to produce regions of p-type conductivity on either side of the remaining region of n-type conductivity.

Electrical connections are made to each of the three different regions as suggested in Fig. 2.2. The thin central layer is known as

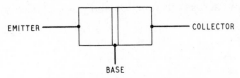

Fig. 2.2. Electrical connections to a bipolar transistor

the *base* of the transistor and corresponds with the control grid of a triode valve. One of the remaining two layers is known as the *emitter* and corresponds with the cathode of a triode valve. The remaining (third) layer is known as the *collector*: it corresponds with the anode of the triode. The transistor may be symmetrical and either of the outer layers may then be used as emitter: the operating conditions determine which of the outer layers behaves as emitter, because in normal operation the emitter-base junction is forward-biased whilst the base-collector junction is reverse-biased. In practice most bipolar transistors are unsymmetrical with the collector junction larger than the emitter junction and it is essential to adhere to the emitter and collector connections prescribed by the manufacturer.

The symbols used for bipolar transistors in circuit diagrams are given in Fig. 2.3. The symbol shown at (a), in which the emitter

(a)   (b)

Fig. 2.3. Circuit diagram symbols for (a) pnp and (b) an npn bipolar transistor

arrow is directed towards the base, is used for a pnp transistor and the symbol shown at (b), in which the emitter arrow is directed away from the base, is used for an npn transistor.

An account of the principal methods used in the manufacture of germanium and silicon transistors is given in Appendix A.

Operation of a pnp Transistor

Fig. 2.4 illustrates the polarity of the potentials which are necessary in a pnp-transistor amplifying circuit. The emitter is biased slightly positively with respect to the base: this is an example of forward bias and the external battery opposes the internal potential barrier associated with the emitter-base junction. A considerable current therefore flows across this junction and this is carried by holes from

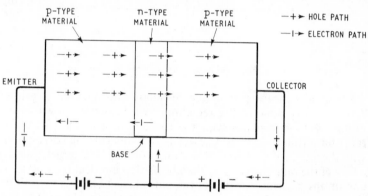

Fig. 2.4. Hole and electron paths in a pnp transistor connected for amplification

the p-type emitter (which move to the right into the base) and by electrons from the n-type base (which move to the left into the emitter). However, because the impurity concentration in the emitter is normally considerably greater than that of the base (this is adjusted during manufacture), the holes carrying the emitter-base current greatly outnumber the electrons and we can say with little error that the current flowing across the emitter-base junction is carried by holes moving from emitter to base. Because holes and electrons play a part in the action, this type of transistor is known as *bipolar*.

The collector is biased negatively with respect to the base: this is an example of reverse bias and the external battery aids the internal potential barrier associated with the base-collector junction. If the emitter-base junction were also reverse-biased, no holes would be injected into the base region from the emitter and only a very small current would flow across the base-collector junction. This is the reverse current (described in Chapter 1): it is a saturation current independent of the collector-base voltage. However, when the emitter-base junction is forward-biased, the injected holes have a marked effect on the collector current: this is the essence of bipolar transistor action. The holes are injected into the base, which is a particularly thin layer; most of them cross the base by diffusion and on reaching the collector-base junction are swept into the collector region. The reverse bias of the base-collector junction ensures the collection of all the holes crossing this junction, whether these are present in the base region as a result of breakdown of covalent bonds by virtue of thermal agitation or are injected into it by the action of the emitter. A few of the holes which leave the emitter combine with electrons in the base and so cease to exist but the majority of the holes (commonly more than 95 per cent) succeed in reaching the collector. Thus the increase in collector current due to hole-injection by the emitter is nearly equal to the current flowing across the emitter-base junction. The balance of the emitter carriers (equal to, say, 5 per cent) is neutralised by electrons in the base region and to maintain charge neutrality more electrons flow into the base, constituting a base current. The collector current, even though it may be considerably increased by forward bias of the emitter-base junction, is still independent of the collector voltage. This is another way of saying that the output resistance of the transistor is extremely high: it can in fact be several megohms. The input resistance is approximately that of a forward-biased junction diode and is commonly of the order of 25 Ω. A small change in the input (emitter) current of the transistor is faithfully reproduced in the output (collector) current but, of course, at a slightly smaller amplitude. Clearly such an amplifier has no current gain but because the output resistance is many times the input resistance it can give voltage gain. To illustrate this suppose a 1-mV signal source is connected to the 25-Ω input. This gives rise to an emitter current of 1/25 mA, i.e. 40 μA. The collector current is slightly less than this but as an approximation suppose the output current is also 40 μA. A common value of load resistance is 5 kΩ and for this value the output voltage is given by $5,000 \times 40 \times 10^{-6}$, i.e. 200 mV, equivalent to a voltage gain of 200.

Bias Supplies for a pnp Transistor

Fig. 2.5 shows a pnp transistor connected to supplies as required in one form of amplifying circuit. For forward bias of the emitter-base junction, the emitter is made positive with respect to the base; for reverse bias of the base-collector junction, the collector is made negative with respect to the base. Fig. 2.5 shows separate batteries used to provide these two bias supplies and it is significant that the batteries are connected in series, the positive terminal of one being connected to the negative terminal of the other. The base voltage in fact lies between that of the collector and the emitter and thus a single battery can be used to provide the two bias supplies by connecting it between emitter and collector, the base being returned to

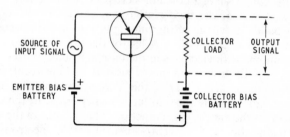

Fig. 2.5. Basic circuit for using a pnp transistor as an amplifier

a tapping point on the battery or to a potential divider connected across the battery. The potential divider technique (Fig. 2.6) is often used in transistor circuits and a pnp transistor operating with the emitter circuit earthed requires a negative collector voltage. The

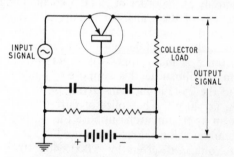

Fig. 2.6. The circuit of Fig. 2.5 using a single battery and a potential divider providing base bias

arrow in the transistor symbol shows the direction of conventional current flow, i.e. is in the opposite direction to that of electron flow through the transistor.

Operation of an npn Transistor

The action of an npn junction transistor is similar to that of a pnp type just described but the bias polarities and directions of current flow are reversed. Thus the charge carriers are predominantly electrons and the collector bias voltage for an earthed-emitter circuit must be positive.

Common-base, Common-emitter and Common-collector Amplifiers

So far we have described amplifying circuits in which the emitter current determines the collector current: it is, however, more usual in transistor circuits to employ the external base current to control the collector or emitter current. Used in this way the transistor is a current amplifier because the collector (and emitter) current can easily be 100 times the controlling (base) current and variations in the input current are faithfully portrayed by much larger variations in the output current.

Thus we can distinguish three ways in which the transistor may be used as an amplifier:

(a) with emitter current controlling collector current,
(b) with base current controlling collector current,
(c) with base current controlling emitter current.

It is significant that in all these modes of use, operation of the transistor is given in terms of input and output current. This is an

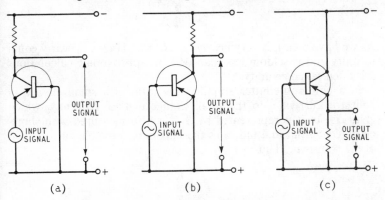

Fig. 2.7. *The three basic forms of transistor amplifier; (a) common-base, (b) common-emitter and (c) common-collector (emitter follower). For simplicity base d.c. bias is omitted*

inevitable consequence of the physics of the bipolar transistor: such transistors are *current-controlled* devices: by contrast thermionic valves and field-effect transistors are voltage-controlled devices.

Corresponding to the three modes of operation listed above there are three fundamental transistor amplifying circuits: these are shown in Fig. 2.7. At signal frequencies the impedance of the collector voltage supply is assumed negligibly small and thus we can say for circuit (a) that the input is applied between emitter and base and that the output is effectively generated between collector and base. Thus the base connection is common to the input and output circuits: this amplifier is therefore known as the *common-base* type.

In (b) the input is again applied between base and emitter but the output is effectively generated between collector and emitter. This is therefore the *common-emitter* amplifier, probably the most used of all transistor amplifying circuits.

In (c) the input is effectively between base and collector, the output being generated between emitter and collector. This is the *common-collector* circuit but it is better known as the *emitter follower*.

Current Amplification Factor

In a common-base amplifier the ratio of a small change in collector current i_c to the small change in emitter current i_e which gives rise to it is known as the current amplification factor α. It is measured with short-circuited output. Thus we have

$$\alpha = \frac{i_c}{i_e} \tag{2.1}$$

As we have seen i_c is very nearly equal to i_e. Thus α is nearly equal to unity and is seldom less than 0·95. In approximate calculations α is often taken as unity.

In a common-emitter amplifier the ratio of a small change in collector current i_c to the small change in base current i_b which gives rise to it is represented by β. It is also measured with short-circuited output and indicates the maximum possible current gain of the transistor. Thus

$$\beta = \frac{i_c}{i_b} \tag{2.2}$$

We have seen how the emitter current of a transistor effectively divides into two components in the base region. Most of it passes into the collector region and emerges as external collector current. The remaining small fraction forms an external base current. This division applies equally to steady and signal-frequency currents. Thus we have

$$i_e = i_c + i_b \tag{2.3}$$

From Eqns 2.1, 2.2 and 2.3 we can deduce a relationship between α and β thus

$$\beta = \frac{i_c}{i_b} = \frac{i_c}{i_e - i_c}$$

But $i_c = \alpha i_e$

$$\therefore \beta = \frac{\alpha i_e}{i_e - \alpha i_e}$$

$$= \frac{\alpha}{1 - \alpha} \tag{2.4}$$

As α is nearly equal to unity there is little error in taking β as given by

$$\beta = \frac{1}{1 - \alpha} \tag{2.5}$$

Thus for a transistor for which $\alpha = 0.98$

$$\beta = \frac{1}{1 - 0.98} = \frac{1}{0.02} = 50$$

In practice values of β lie between 20 and 500. β is one of the properties of a transistor normally quoted in manufacturer's literature: here it is usually known as h_{fe}. Values of β for silicon transistors tend to be higher than for germanium.

Collector Current—Collector Voltage Characteristics

Fig. 2.8(a) illustrates the way in which the collector current of a bipolar transistor varies with collector voltage for given values of emitter current. The curves illustrate the point made immediately above namely that the collector current is always slightly less than the emitter current.

The characteristics have a shape similar to those of the anode current—anode voltage curves for a pentode, the near horizontal portions indicating the high collector a.c. resistance of the transistor. However, they represent a better approximation to the ideal characteristic than those of a pentode because the transistor curves have no knee to limit the swing of collector during amplification. The collector voltage can swing the whole extent of the collector supply voltage, giving the transistor an efficiency equal to the theoretical maximum.

Fig. 2.8(b) illustrates the way in which the collector current of a bipolar transistor varies with collector voltage for given values

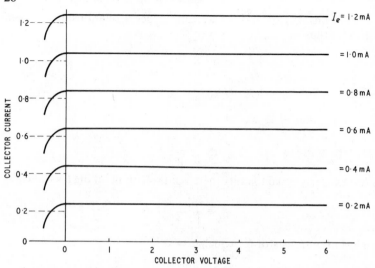

Fig. 2.8(a) A set of $I_c - V_c$ characteristics for a common-base amplifier

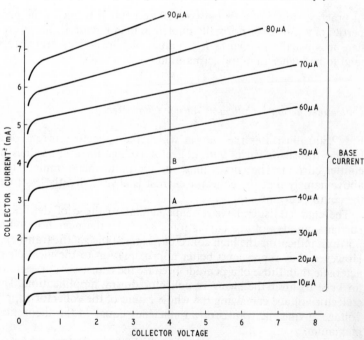

Fig. 2.8(b) Typical collector current–collector voltage characteristics for common-emitter connection

of base current. These are also similar in shape to the anode current-
anode voltage curves for a pentode but the knee of the characteristics
occurs at a very low collector voltage, permitting large swings in
collector voltage during amplification.

A number of parameters of the transistor can be obtained from
these characteristics. For example the slope of the curves is not so
low as for the common-base connection showing that the collector
a.c. resistance is smaller: it is in fact approximately 30 kΩ. To deduce
β consider the intercepts made by the characteristics on the vertical
line drawn through $V_c = 4$ V. When the base current is 40 μA (point
A) the collector current is 2·5 mA and when I_b is 50 μA (point B)
I_c is 3·5 mA. A change of base current of 10 μA thus causes a change
in collector current of 1 mA: this corresponds to a value of β of 100.

Collector Current—Base Voltage Characteristics

It is useful to consider the $I_c - V_{be}$ curves for transistors because
these illustrate one of the important differences between germanium
and silicon transistors. Typical curves are given in Fig. 2.9 and these
show that the germanium transistor takes a useful collector current
for very small base-emitter voltages: in fact it is common practice

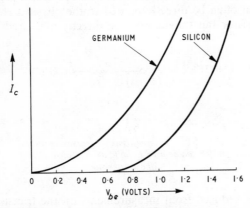

Fig. 2.9. $I_c - V_{be}$ *curves for germanium and silicon transistors*

in approximate calculations to assume the base-emitter voltage to
be zero for normal amplifying operation. For silicon transistors,
however, the base-emitter voltage must be a good fraction of a volt,
commonly 0·7 V, to give a useful collector current: this is commonly
referred to as the greater offset of the silicon transistor.

Equivalent Circuit of a Transistor

For calculating the performance of transistor circuits, it is useful
to regard the transistor as a three-terminal network which is specified
in terms of its input resistance, output resistance and current gain—
all fundamental properties which can readily be measured. The
properties of such a network are expressed in a number of ways
notably as z parameters, y parameters or h parameters. The basic
equations for these parameters are given in Appendix B and this
shows that these parameters, in spite of their variety, express the
same four properties of the network, namely its input resistance,
output resistance, forward gain and reverse gain.

One of the disadvantages of this method of expressing transistor
properties is that the values of the fundamental properties which
apply to the common-base connection, do not apply to the common-
emitter connection or to the common-collector connection and three
sets of values are therefore required in a complete expression of a
transistor's properties. Moreover the numerical values of these properties vary with emitter current and frequency and can be regarded
as constant only over a narrow frequency and emitter-current range.

An alternative approach to the problem of calculating transistor
performance is to deduce an equivalent network which has a behaviour similar to that of the transistor. The constants of such a
network can often be directly related to the physical construction
of the transistor but they cannot be directly measured: they can,

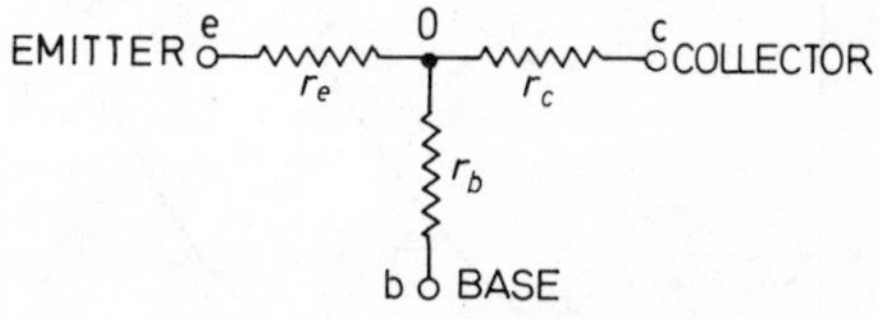

*Fig. 2.10. A three-terminal passive network
which can be used to build an equivalent circuit
for a transistor*

however, be deduced from measurements on the transistor. If the
network is truly equivalent it will hold at all frequencies and by
applying Kirchhoff's laws or other network theorems to this equivalent circuit we can calculate the performance of the circuit. Much
useful work is possible by representing a transistor as a simple
T-network of resistance as shown in Fig. 2.10.

The transistor cannot, however, be perfectly represented by three
resistances only because such a network cannot generate power

(as a transistor can) but can only dissipate power. In other words the network illustrated in Fig. 2.10 is a *passive* network and to be accurate the equivalent network must include a source of power, i.e. must be *active*.

The source of power could be shown as a constant-current source connected in parallel with the collector resistance r_c as in Fig. 2.11(a)

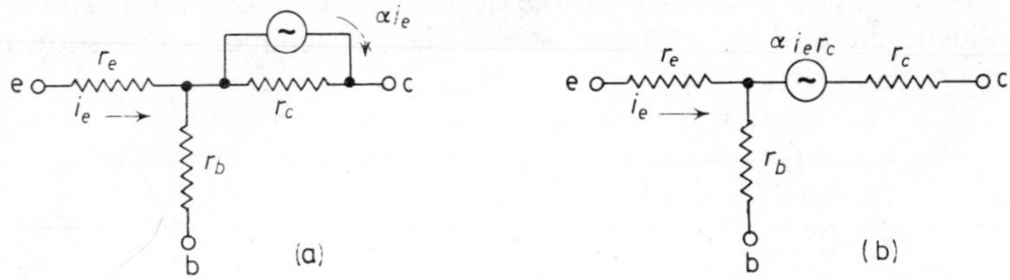

Fig. 2.11. A three-terminal active network which can be used as an equivalent circuit for a transistor (a) including a constant-current generator, and (b) a constant-voltage generator

and the current so supplied is, as we have already seen, equal to αi_e, where α is the current amplification factor of the transistor and i_e is the current in the emitter circuit, i.e. in the emitter resistance r_e. Alternatively the source of power can be represented as a constant-voltage generator connected in series with r_c as shown in Fig. 2.11(b).

Such a voltage has precisely the same effect as the constant-current generator, provided the voltage is given the correct value, and the value required is equal to $\alpha i_e r_c$ as can be shown by applying Thevenin's theorem to Fig. 2.11(a).

In some books on transistors the voltage generator is given as $r_m i_e$, where r_m is known as the mutual or transfer resistance* and is equal to αr_c. The mutual resistance may be defined as the ratio of the e.m.f. in the collector circuit to the signal current in the emitter circuit which gives rise to it. This may be regarded as the dual of the mutual conductance g_m of a thermionic valve which is defined as the ratio of the current in the anode circuit to the signal voltage in the grid circuit which causes it.

This comparison is useful because it again reminds us that the transistor is a device controlled by an input current rather than an input voltage.

This network is quite satisfactory for calculating the performance of transistor amplifiers at low frequencies such as audio amplifiers because the reactances of the internal capacitances have, in general, negligible effects on performance.

* The word 'transistor' is, in fact, derived from 'transfer resistance'.

At higher frequencies, however, and in particular at radio frequencies, it is necessary to include such capacitances in the T-network to obtain accurate answers.

Of the various internal capacitances within a bipolar transistor, that between the collector and the base has the greatest effect on the high-frequency performance.

In a transistor r.f. amplifier the capacitance between collector and base provides feedback from the output to the input circuit. This is illustrated in Fig. 2.12(a) in which the capacitance c_{bc} is shown

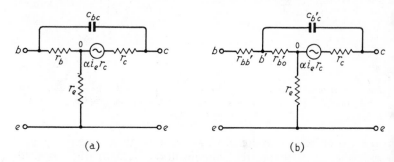

Fig. 2.12. *T-section equivalent circuit of a transistor showing collector capacitance, (a) returned to base input terminal and (b) returned to a tapping point on the base resistance*

connected directly between collector and base terminals. However, a better approximation to the performance of a transistor at high frequencies is obtained by assuming that the collector capacitance is returned to a tapping point b' on the base resistance as shown in Fig. 2.12(b).

This modifies the feedback which now occurs via $c_{b'c}$ and $r_{bb'}$ in series and the high-frequency performance of the transistor is now dependent on the time constant $r_{bb'}\,c_{b'c}$ which is probably the most important characteristic of a transistor intended for high-frequency use.

Values of r_e, r_b *and* r_c

The networks of Fig. 2.11 can be used in analyses of transistor circuits. The resistances represent differential, i.e. a.c. quantities and, for a given value of mean emitter current I_e, are constant provided that the variations in voltage or current which occur during operation of the transistor are small.

The value of r_e for all transistors, irrespective of size and type, is given by

$$r_e = \frac{kT}{eI_e}$$

where k = Boltzmann's constant, i.e.$1 \cdot 374 \times 10^{-23}$ J/°C

e = charge on the electron, i.e. $1 \cdot 59 \times 10^{-19}$ C

and T = absolute temperature.

Thus r_e is directly proportional to the absolute temperature and inversely proportional to the emitter current. The coefficient kT/e has the dimensions of a voltage and, if we substitute numerical values for k and e this voltage is 25 mV at a temperature of 20°C. Thus we have

$$r_e\ (\Omega) = \frac{25\ (mV)}{I_e\ (mA)}$$

If I_e = 1 mA, r_e is 25 Ω (at 20°C).

The value of r_c is also inversely proportional to I_e in simple transistors but the value depends on the impurity grading in the base region and is very high for diffused-base transistors.

The value of r_b is largely independent of I_e and is a function of the size and geometry of the transistor. For small alloy-junction transistors r_b can be as high as 500 Ω, for a silicon mesa transistor it is unlikely to exceed 100 Ω but for a large power transistor r_b may be as low as 5 Ω.

The values of r_e, r_b and r_c cannot be measured directly because it is impossible to obtain a connection to the point O (in Fig. 2.12) but the values can be deduced from measurements made at the transistor terminals.

Although the bipolar transistor is a current-controlled device there are occasions (e.g. when it is used as a voltage amplifier) when it is useful to be able to express the performance in terms of the signal voltage applied between base and emitter. For such calculations we need to know the mutual conductance g_m of the transistor: this is the ratio of a small change in collector current to the change in base-emitter voltage which causes it.

The mutual conductance of a transistor is given by

$$g_m = \frac{\alpha_o}{r_e}$$

where α_o is the current amplification factor at low frequencies. α_o is very nearly equal to unity and thus we may say

$$g_m \approx \frac{1}{r_e}$$

We know, from page 33, that r_e is inversely proportional to the mean emitter current I_e according to the relationship

$$r_e\,(\Omega) = \frac{25\,(\text{mV})}{I_e\,(\text{mA})}$$

Eliminating r_e between the last two expressions we have

$$g_m\,(\text{A/V}) \approx \frac{I_e\,(\text{mA})}{25\,(\text{mV})}$$

which is perhaps more conveniently expressed

$$g_m\,(\text{mA/V}) \approx 40 I_e(\text{mA})$$

Thus for an emitter current of 1 mA the mutual conductance is 40 mA/V.

Frequency f_T

Modern circuitry makes extensive use of diffused-based, e.g. silicon planar transistors as common-emitter amplifiers and the high-frequency performance is usually expressed in terms of the parameter f_T, the transition frequency. This is defined as the frequency at which the modulus of β (the current gain of the common-emitter amplifier) has fallen to unity. It thus measures the highest frequency at which the transistor can be used as an amplifier: it also gives the gain-bandwidth product for the transistor.

FIELD-EFFECT TRANSISTORS

Introduction

Field-effect transistors operate on principles quite different from those of bipolar transistors. They consist essentially of a channel of semiconducting material, the charge-carrier density of which is controlled by the input signal. The input signal thus determines the conductivity of the channel and hence the current which flows through it from the supply. The connections from the ends of the channel to the supply are known as the source and drain terminals and they correspond to the emitter and collector terminals of a bipolar transistor. The control terminal is known as the gate. The potential on the gate can control the channel conductivity via a reverse-biased pn junction: transistors using this principle are termed junction-gate field-effect transistors (j.u.g.f.e.t.s). Alternatively the gate potential can control the channel conductivity via a capacitance link: transistors using this principle are termed insulated-gate field-effect transistors (i.g.f.e.t.s). F.e.t.s in general can be regarded as voltage-controlled variable resistors.

There is only one type of charge carrier in an f.e.t. namely electrons in an n-channel device and holes in a p-channel device. Thus f.e.t.s can be termed unipolar* transistors: those previously described in this chapter are, of course, bipolar. An important consequence of the single type of charge carrier is that an f.e.t. can introduce less noise than bipolar types.

Junction-gate Field-effect Transistors (j.u.g.f.e.t.s)

The structure of an n-channel j.u.g.f.e.t. is illustrated in Fig. 2.13. This shows a slab of high-resistance n-type silicon with ohmic contacts formed by highly doped (n+) regions near the two ends: these provide the source and drain connections to the external

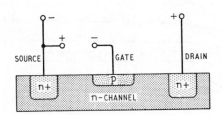

Fig. 2.13. Structure of an n-channel j.u.g.f.e.t. with bias polarities indicated

circuit. A region of p-type conductivity is formed between the ohmic contacts: this forms the gate connection and is reverse-biased, i.e. negatively biased with respect to the source connection. When the source and drain terminals are connected to a supply, current flows longitudinally through the slab, being carried by the free electrons of the n-type material. However, near the p-region there is a depletion

---

* Strictly this should be monopolar if we are using Greek prefixes.

area, i.e. an area free of charge carriers and if the reverse bias of the gate is increased the depletion area spreads, confining the longitudinal current to a small cross-sectional area of the slab and reducing its amplitude. If the reverse bias is increased sufficiently the depletion area spreads to the whole cross-section of the slab and cuts off the current completely. The gate-source voltage required to do this is known as the pinch-off voltage. By connecting a signal source in series with the gate bias the effective width of the channel can be modulated and the signal waveform is impressed on the current. If the current is passed through a suitable external impedance, the voltage generated across it is a magnified version of that of the signal source.

The input resistance of a j.u.g.f.e.t. is that of a reverse-biased pn junction and can be very high: a typical value is 10^{10} Ω with a shunt capacitance of say 5 pF. These values are more like those of a thermionic valve than a bipolar transistor. The characteristics of a j.u.g.f.e.t. are similar to those of a pentode valve as shown in the

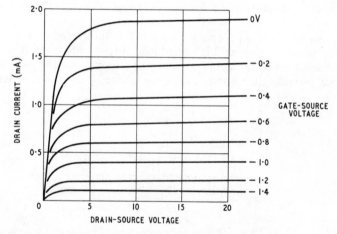

Fig. 2.14. *Typical characteristics for a field-effect transistor*

typical example of Fig. 2.14: these are plotted in terms of gate-source voltage because the field-effect transistor is voltage-controlled. If the gate terminal of the transistor illustrated in Fig. 2.13 is biased positively with respect to the source, the pn junction is forward biased giving a very low input resistance: yet another point of similarity with a thermionic valve.

There is a complementary type of j.u.g.f.e.t. with a p-type channel and n-type gate: this requires a positive gate-source voltage to cut off the channel current. The graphical symbols for both types of j.u.g.f.e.t. are given in Fig. 2.15.

The j.u.g.f.e.t. takes a significant drain current with zero gate bias and, for an n-channel device, a negative bias is required to cut the current off. When the f.e.t. is used as an amplifier the gate bias is normally between zero and cut off and lies outside the range of the drain-source voltage; for example typical voltages are $V_g = -1$ V, $V_s = 0$ and $V_d = +20$ V. The grid bias for thermionic valves similarly

Fig. 2.15. *Graphical symbols for j.u.g.f.e.t.s*

lies outside the anode-cathode potential range. Devices with this property are said to operate in the *depletion mode*. The base bias voltage for a bipolar transistor lies between that of the emitter and the collector.

Insulated-gate Field-effect Transistors (i.g.f.e.t.s)

A cross-section of an i.g.f.e.t. is shown in Fig. 2.16. It consists of a base (substrate) layer of p-type silicon into the surface of which two

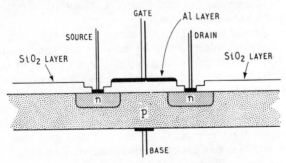

Fig. 2.16. *Structure of enhancement-type n-channel i.g.f.e.t.*

closely spaced stripes of n-type conductivity are diffused. Ohmic connections are made to the n regions to give source and drain connections, and to the p-type substrate to give a base connection. The device is sealed by a silicon dioxide coating and a thin layer of aluminium which provides a gate connection. The device can be manufactured by the techniques used for planar transistors described in Appendix A.

The substrate, the silicon-dioxide layer and the aluminium skin form a parallel-plate capacitor. The voltage applied between gate and base controls the conductivity of the surface area of the substrate between the n regions and hence the current which flows between source and drain when these are connected to a supply. If the gate-base voltage is zero the only drain current is the leakage current of one of the pn junctions and this, in a silicon device, is very small indeed. If the gate is made positive with respect to the base, positive charge carriers are repelled into the body of the substrate and negative charge carriers are attracted to its surface either from thermal breakdown of the p material or from the n regions. In this way a layer of mobile charge carriers is induced on the surface of the substrate and this makes ohmic contact with the diffused n regions. The induced layer provides the conducting channel between the source and drain terminals and permits drain current to flow. The induced layer is known as an inversion layer because it changes its conductivity from p-type to n-type as the gate voltage is increased positively from zero. Increase in gate voltage increases the number of charge carriers in the induced layer, increasing channel conductivity and drain current. The gate voltage thus controls the drain current and the characteristics have a form similar to those of a j.u.g.f.e.t. A significant feature of the behaviour of this type of i.g.f.e.t. is that drain current is zero in the absence of a gate voltage and that a forward (positive for an n-channel device) gate bias is necessary to give working values of drain current: this type of operation is known as the *enhancement mode.*

For an enhancement i.g.f.e.t. to obtain a working value of drain current, the gate voltage must lie between the source and drain potentials. This compares with bipolar transistors and contrasts with depletion devices such as j.u.g.f.e.t.s and thermionic valves.

It is, however, possible in the manufacture of an i.g.f.e.t. to produce a thin n-type layer on the surface of the p-type substrate. This provides a conducting channel between source and drain and ensures that a useful drain current flows even at zero gate-base voltage. Negative gate bias reduces channel conductivity and drain current and in the limit will change the conductivity of the channel to p type, reducing drain current to zero: this is the depletion mode of operation again. Positive voltages on the gate increase channel conductivity and drain current as in enhancement-mode operation. I.g.f.e.t.s of this type can thus operate in depletion and enhancement modes.

Complementary i.g.f.e.t.s with an n-type substrate and p-type channel are also available, giving a total of four basic types of i.g.f.e.t.: the graphical symbols are shown in Fig. 2.17. The non-conductivity of the enhancement type for zero gate bias is indicated in the symbol by breaks in the rectangle representing the channel.

Input resistances greater than 10^{12} Ω have been achieved in i.g.f.e.t.s and input capacitances may be as low as 1 pF.

The connection to the substrate provides a second input terminal (the base) to an i.g.f.e.t. and the potential applied to it with respect to the source controls the drain current in the same way as a potential

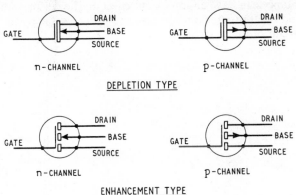

Fig. 2.17. *Graphical symbols for i.g.f.e.t.s.*

on the gate of a j.u.g.f.e.t. The base terminal is not so sensitive a control electrode as the gate terminal but is used in certain types of circuit (see p. 209). In many i.g.f.e.t. applications the base terminal is connected to the source terminal internally or externally.

The silicon-dioxide dielectric of the i.g.f.e.t. is very thin and can easily be broken down by transient voltages on the gate terminal. A practical precaution normally taken in handling such transistors is to keep the gate terminal connected to the base whilst the transistor is being connected into circuit.

Equivalent Circuit for an f.e.t.

The full equivalent circuit for an f.e.t. is complex but at low frequencies the effects of internal capacitances can usually be ignored and the input resistance can often be taken as infinite. The circuit then simplifies to the form shown in Fig. 2.18(a).

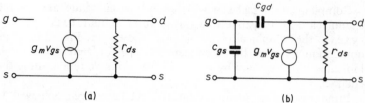

Fig. 2.18. Equivalent circuit for an f.e.t. for use at (a) low frequencies and (b) radio frequencies

At radio frequencies the input capacitance and the gate-drain capacitance can have an important effect on the performance of the f.e.t. and the transistor is more accurately represented by the equivalent circuit of Fig. 2.18(b).

CHAPTER 3

Common-base and Common-gate Amplifiers

COMMON-BASE AMPLIFIERS

Introduction

It was shown in the previous chapter that the common-base amplifier has a current gain of just less than unity, that its input resistance is low, output resistance high and that it can give substantial voltage gain. In this chapter we shall show how these resistances and the

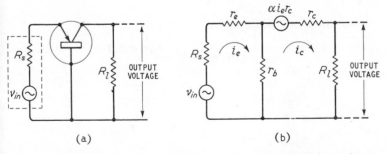

Fig. 3.1. Simplified circuit of a common-base amplifier (a), and its electrical equivalent (b)

voltage gain depend on the values of r_e, r_b and r_c and on the value of the source and load resistance.

A simplified circuit of a transistor common-base amplifier is given in Fig. 3.1(a) in which R_s represents the internal resistance of the signal source and R_l represents the collector load for the amplifier.

41

For the sake of simplicity, methods of applying bias are not indicated.

Fig. 3.1(b) represents the same circuit in which the transistor is represented by its equivalent T-section network. The directions of the currents in the two meshes of this network can be determined by considering the physical processes occurring in a transistor. From Chapter 2 we know that the emitter current i_e is equal to the sum of the collector current i_c and the base current i_b. If, therefore, i_e and i_c are given clockwise directions as shown in Fig. 3.1(b), then the current in r_b is $(i_e - i_c)$ which is equal to i_b: all conventions are therefore satisfied in this diagram.

Input Resistance

Applying Kirchhoff's laws to the circuit of Fig. 3.1(b) we have

$$v_{in} = i_e(R_s + r_e + r_b) - i_c r_b \qquad (3.1)$$

$$0 = i_c(r_b + r_c + R_l) - i_e r_b - \alpha i_e r_c \qquad (3.2)$$

To obtain an expression for the input resistance we can eliminate i_c between these two equations to obtain a relationship between i_e and v_{in}.

From Eqn 3.2

$$i_c = \frac{r_b + \alpha r_c}{r_b + r_c + R_l} i_e$$

Substituting for i_c in Eqn 3.1

$$v_{in} = i_e(R_s + r_e + r_b) - \frac{r_b(r_b + \alpha r_c)}{r_b + r_c + R_l} i_e$$

$$\therefore i_e = \frac{v_{in}}{R_s + r_e + r_b - \dfrac{r_b(r_b + \alpha r_c)}{r_b + r_c + R_l}} \qquad (3.3)$$

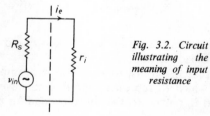

Fig. 3.2. Circuit illustrating the meaning of input resistance

If we represent the input resistance of the transistor by r_i the circuit of Fig. 3.1(b) takes the form shown in Fig. 3.2. For this circuit we have

$$i_e = \frac{v_{in}}{R_s + r_i} \tag{3.4}$$

Comparing Eqns 3.3 and 3.4

$$r_i = r_e + r_b - \frac{r_b(r_b + \alpha r_c)}{r_b + r_c + R_l} \tag{3.5}$$

$$= r_e + r_b \cdot \frac{R_l + r_c(1 - \alpha)}{r_b + r_c + R_l} \tag{3.6}$$

For a given transistor operating under constant d.c. conditions, r_b, r_e, r_c and α are constant, and r_i therefore depends solely on R_l, increasing as R_l increases.

The range of input resistance can be calculated from Eqn 3.5. First consider Eqn 3.5 when R_l is made vanishingly small

$$r_i = r_e + r_b - \frac{r_b(r_b + \alpha r_c)}{r_b + r_c}$$

Now for bipolar transistors r_c and αr_c are both large compared with r_b and the input resistance is approximately given by

$$r_i = r_e + r_b - r_b \cdot \frac{\alpha r_c}{r_c}$$

$$= r_e + r_b(1 - \alpha)$$

From Eqn 2.5 $\qquad (1 - \alpha) = \frac{1}{\beta}$

$$\therefore r_i = r_e + \frac{r_b}{\beta} \tag{3.7}$$

and this gives the value of the input resistance of the common-base amplifier when the output terminals are short-circuited.

Frequently r_b/β can be neglected in comparison with r_e and thus we can say

$$r_i \approx r_e$$

As we have seen r_e is inversely proportional to emitter current. It follows that the input resistance for the common-base amplifier (for low values of collector load resistance) is inversely proportional to emitter current and thus to collector current. This has important consequences. For example suppose a series of steady potentials is applied between base and emitter as in plotting the $I_c - V_{be}$ curves for the transistor. Each time the voltage is increased, the emitter current increases and r_e decreases with the result that the final increase in I_e is more than proportional to the voltage increase. Similarly if V_{be} is decreased I_e decreases and r_e increases, the effect being that the final decrease in I_e is less than proportional to the change in voltage. Thus the $I_c - V_{be}$ curve is far from linear: indeed it approximates to exponential form as shown in Fig. 2.9.

Changes in input resistance also occur when emitter current varies during signal amplification in the transistor. If the input signal is from a high-resistance source then variations in input resistance have little effect on input current and amplification is linear. If however the transistor is used as a voltage amplifier and the input signal is from a low-resistance source, variations in input resistance can cause severe distortion; this is to be expected from a device with a characteristic as non-linear as that shown in Fig. 2.9. To minimise this distortion the effects of the input-resistance variation must be eliminated and one way of achieving this is to ensure that the internal resistance of the signal source is large compared with the average value of the input resistance of the transistor. If this is impossible and the source resistance is small, it can effectively be increased by connecting an external resistance in series with the signal source. By making R_s large we ensure that the transistor is fed with a current that is a substantially undistorted copy of the voltage to be amplified: thus we still use the transistor as a current amplifier even though the input and output signals are in the form of voltages.

Now consider the input resistance when R_l is made very large. The final term in Eqn 3.5 vanishes leaving

$$r_i = r_e + r_b \qquad (3.8)$$

which gives the value of the input resistance when the output terminals are open-circuited.

Input Resistance for a Bipolar Transistor as Common-base Amplifier

To obtain an estimate of practical values of input resistance for a bipolar transistor we can substitute $r_e = 25\ \Omega$, $r_b = 300\ \Omega$ and $\beta = 50$ and from Eqn 3.7 we find that the input resistance for short-circuited output terminals is given by

$$r_i = 25 + \frac{300}{50}$$

$$= 31 \ \Omega$$

From Eqn 3.8 the input resistance for open-circuited output terminals is given by

$$r_i = 25 + 300$$

$$= 325 \ \Omega$$

Thus the input resistance of a bipolar transistor used as a common-base amplifier varies between a minimum value for

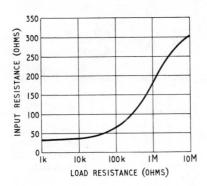

Fig. 3.3 Variation of input resistance with load resistance for a common-base amplifier

short-circuited output terminals to a maximum value for open-circuited terminals. The dependence of input resistance on output load is illustrated in Fig. 3.3.

Output Resistance

To obtain an expression for the output resistance of the common-base amplifier we can eliminate i_e between Eqns 3.1 and 3.2 to obtain a relationship between i_c and v_{in}.

From Eqn 3.2

$$i_e = \frac{r_b + r_c + R_l}{r_b + \alpha r_c} \cdot i_c$$

Substituting for i_e in Eqn 3.1

$$v_{in} = \frac{(r_b + r_c + R_l)(R_s + r_e + r_b)}{r_b + \alpha r_c} i_c - i_c r_b$$

$$\therefore i_c = \frac{v_{in}}{\dfrac{(r_b + r_c + R_l)(R_s + r_e + r_b)}{r_b + \alpha r_c} - r_b}$$

$$= \frac{(r_b + \alpha r_c) v_{in}}{(r_b + r_c + R_l)(R_s + r_e + r_b) - r_b(r_b + \alpha r_c)}$$

$$\therefore i_c = \frac{\dfrac{r_b + \alpha r_c}{R_s + r_e + r_b} v_{in}}{R_l + r_b + r_c - \dfrac{r_b(r_b + \alpha r_c)}{R_s + r_e + r_b}} \tag{3.9}$$

If we represent the output resistance of the transistor amplifier

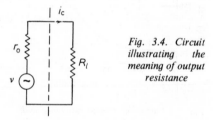

Fig. 3.4. Circuit illustrating the meaning of output resistance

by r_o the circuit of Fig. 3.1(b) has the form shown in Fig. 3.4. For this circuit we have

$$i_c = \frac{v}{R_l + r_o} \tag{3.10}$$

Comparing Eqns 3.9 and 3.10 we have that v, the e.m.f. effectively acting in the output circuit, is given by

$$v = \frac{r_b + \alpha r_c}{R_s + r_e + r_b} v_{in}.$$

Making the normally-justified assumptions that r_b is negligible compared with αr_c, that α can be taken as unity and that $(r_e + r_b)$ can be neglected in comparison with R_s we have

$$v \approx \frac{r_c}{R_s} . v_{in}$$

Eqn 3.10 thus becomes

$$i_c = \frac{\dfrac{r_c}{R_s} \cdot v_{in}}{R_l + r_o}$$

Comparing Eqns 3.9 and 3.10, we have

$$r_o = r_b + r_c - \frac{r_b(r_b + \alpha r_c)}{R_s + r_e + r_b} \tag{3.11}$$

$$= r_c + r_b \cdot \frac{R_s + r_e - r_c\alpha}{R_s + r_e + r_b} \tag{3.12}$$

For a given transistor operating with fixed d.c. conditions, r_b, r_e, r_c and α are fixed and r_o then depends on R_s, increasing as R_s is increased. The range over which r_o varies can be estimated as follows. First let $R_s = 0$. Eqn 3.12 then becomes

$$r_o = r_c + r_b \cdot \frac{r_e - r_c\alpha}{r_e + r_b}$$

which, if the relatively small term $r_b r_e/(r_e + r_b)$ is neglected, gives

$$r_o = r_c \cdot \frac{r_e + r_b(1 - \alpha)}{r_e + r_b} \tag{3.13}$$

Now let R_s approach infinity. The final term in Eqn 3.11 then vanishes, leaving the output resistance as

$$r_o = r_c + r_b \tag{3.14}$$

$$\approx r_c$$

because r_b is small compared with r_c.

Output Resistance of a Bipolar Transistor as Common-base Amplifier

To obtain a numerical estimate of the range of output resistance for a bipolar transistor let $r_e = 25 \ \Omega$, $r_b = 300 \ \Omega$, $r_c = 1 \ \text{M}\Omega$ and $\alpha = 0.98$. Substituting these values in Eqn 3.13 we find the output resistance for short-circuited input terminals is given by

$$r_o = 1{,}000{,}000 \times \left[\frac{25 + 300 \ (1 - 0.98)}{25 + 300} \right] \Omega$$

$$= 100 \ \text{k}\Omega \ \text{approximately}$$

The output resistance for open-circuited input terminals is, from Eqn 3.14, given by

$$r_o = 1,000,000 + 300 \ \Omega$$

$$= 1,000 \ \text{k}\Omega \ \text{approximately}$$

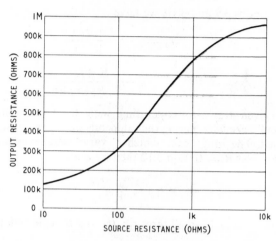

Fig. 3.5. Variation of output resistance with source resistance for a common-base amplifier

The variation of output resistance with source resistance for a common-base amplifier is illustrated in Fig. 3.5.

Voltage Gain

From Fig. 3.1(b) we can see that the output voltage is given by $i_c R_l$. The voltage gain v_{out}/v_{in} is hence equal to $i_c R_l/v_{in}$. From Eqn 3.9 i_c is given by

$$i_c = \frac{(r_b + \alpha r_c)v_{in}}{(R_l + r_b + r_c)(R_s + r_e + r_b) - r_b(r_b + \alpha r_c)}$$

Hence

$$\frac{v_{out}}{v_{in}} = \frac{i_c R_l}{v_{in}} = \frac{(r_b + \alpha r_c) R_l}{(R_l + r_b + r_c)(R_s + r_e + r_b) - r_b(r_b + \alpha r_c)} \quad (3.15)$$

The dependence of the voltage gain on the value of R_l is illustrated in Fig. 3.6, which shows gain increasing with R_l, linearly for small

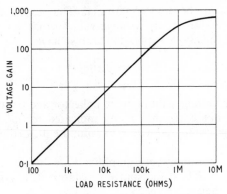

*Fig. 3.6. Variation of voltage gain with load
resistance for a common-base amplifier*

values of R_l but becoming asymptotic to a limiting value of gain
as R_l becomes large.

For a bipolar transistor r_c is normally large compared with all
other resistances in the expression, and it is possible to simplify
Eqn 3.15 as follows:

$$\frac{v_{out}}{v_{in}} = \frac{\alpha r_c R_l}{r_c\,(R_s + r_e + r_b) - \alpha r_b r_c}$$

$$= \frac{\alpha R_l}{R_s + r_e + r_b\,(1 - \alpha)}$$

If α in the numerator is taken as unity, this may be written

$$\frac{v_{out}}{v_{in}} = \frac{R_l}{R_s + r_e + r_b/\beta}$$

From Eqn 3.7

$$r_i = r_e + \frac{r_c}{\beta}.$$

$$\therefore \frac{v_{out}}{v_{in}} = \frac{R_l}{R_s + r_i} \tag{3.16}$$

Now v_{in} is the voltage of the input-signal generator and in a practical
circuit the terminals of v_{in} are not normally accessible. It is more
useful therefore to express the voltage gain of the amplifier as the
ratio of the signal voltage at the collector to that at the emitter
because these two voltages can readily be measured. Reference to
Fig. 3.1. shows that if we make R_s equal to zero, v_{in} equals v_{be}, the

signal voltage at the emitter terminal. Thus to obtain this more practical expression for the voltage gain all that is necessary is to put $R_s = 0$ in the above expression and we have

$$\frac{v_{out}}{v_{be}} = \frac{R_l}{r_i}$$

This gives the voltage gain between collector and emitter terminals and R_s does not enter into it. As indicated earlier, however, R_s should be large compared with r_i to minimise distortion in a class-A amplifier.

In a typical common-base amplifier $R_l = 10 \ \text{k}\Omega$ and $r_i = 40 \ \Omega$ giving v_{out}/v_{be} as 250. If the source resistance R_s is taken as 400 Ω (ten times r_{in}) then v_{out}/v_{in} is 25.

To obtain an indication of the maximum voltage gain available from a bipolar transistor as a common-base amplifier suppose R_l is made large compared with r_c. With a few simplifications Eqn 3.15 reduces to the form

$$\frac{v_{out}}{v_{in}} = \frac{r_c}{R_s + r_e + r_b} \tag{3.17}$$

from which

$$\frac{v_{out}}{v_{be}} = \frac{r_c}{r_e + r_b}.$$

Substituting $r_c = 1 \ \text{M}\Omega$, $r_e = 25 \ \Omega$ and $r_b = 300 \ \Omega$ we have that the maximum value of v_{out}/v_{be} is given by $10^6/(25 + 300)$, approximately 3,000, a much larger value than is obtainable from an f.e.t.

Applications of Common-base Amplifiers

The common-base amplifier has a very low input resistance, a very high output resistance, high voltage gain and unity current gain. It is widely used as an r.f. stage in v.h.f. and u.h.f. receivers, often as part of a cascode stage. Typical circuits are given in later chapters.

COMMON-GATE AMPLIFIERS

Amplifier Properties

Fig. 3.7(a) shows the basic form of the common-gate amplifier and Fig. 3.7(b) gives the (low-frequency) equivalent circuit for an f.e.t.

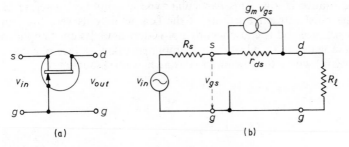

Fig. 3.7. *Basic connections for a common-gate amplifier (a) and the electrical equivalent with source and load circuit (b)*

connected to a signal source and to a load resistor. Because of the very high resistance of the gate circuit, the gate connection is shown as an open circuit in Fig. 3.7(b). The output of the constant-current generator $g_m v_{gs}$ divides between r_{ds} and $(R_s + R_l)$ but normally $(R_s + R_l)$ is small compared with r_{ds} and it can thus be assumed that the current $g_m v_{gs}$ flows wholly through R_s and R_l.

The input resistance of the common-gate amplifier r_i is the ratio of the voltage across the source and gate terminals (v_{gs}) to the current flowing through these terminals ($g_m v_{gs}$) and is thus equal to $1/g_m$—a very low resistance. If $g_m = 2$ mA/V, a typical value, r_i is $1/(2 \times 10^{-3})$, i.e. 500 Ω.

The output resistance r_o can be seen from inspection of Fig. 3.7(b) to be $(r_{ds} + R_s)$ but as r_{ds} is normally large compared with R_s, r_o can usually be taken as approximately equal to r_{ds}.

The output voltage v_o of the common-gate amplifier can be seen from inspection of Fig. 3.7(b) to be iR_l where i is the current ($g_m v_{gs}$) in the circuit. Thus the voltage gain is given by

$$\frac{v_o}{v_{gs}} = g_m R_l$$

The current in the circuit ($g_m v_{gs}$) flows in the input and output circuits and the current gain of the common-gate amplifier is therefore unity.

Applications

The common-gate amplifier has output resistance and voltage gain similar to those of the common-source amplifier but its input resistance is low. This particular type of amplifier therefore lacks the most attractive feature of the f.e.t. namely its very high input resistance and is consequently not widely used. Its chief application is in wideband v.h.f. and u.h.f. amplifiers (often as part of a cascode circuit) where the low input resistance is no disadvantage.

CHAPTER 4

Common-emitter and Common-source Amplifiers

Introduction

We have already seen (page 27) that the common-emitter amplifier has a considerable current gain β. In the following analysis we shall derive expressions for the input resistance, output resistance and voltage gain for this circuit in terms of the parameters of the equivalent T-network, the source resistance R_s and the output load R_l.

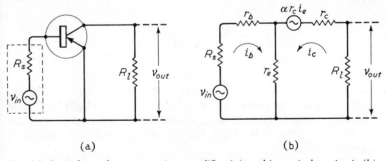

(a)   (b)

Fig. 4.1. Basic form of common-emitter amplifier (a), and its equivalent circuit (b)

The basic circuit for a common-emitter amplifier is given in Fig. 4.1(a), in which R_s is the resistance of the signal source and R_l is the collector load resistance. Bias sources are omitted from this diagram for the sake of simplicity. Fig. 4.1(b) gives the equivalent circuit in which the transistor is represented by the T-network of resistances r_e, r_b and r_c introduced in Chapter 2. If the base current i_b is shown acting in a clockwise direction and the collector current i_c in an anticlockwise direction as indicated in Fig. 4.1(b), then the current in the common-emitter resistance r_e is the sum $(i_b + i_c)$. From Chapter 2 we know that the sum of these two currents is the emitter current i_e: all conventions are therefore satisfied in this diagram.

Applying Kirchhoff's laws to the circuit of Fig. 4.1(b) we have

$$v_{in} = i_b (R_s + r_b + r_e) + i_c r_e \qquad (4.1)$$

$$0 = i_c (R_l + r_c + r_e) + i_b r_e - \alpha r_c i_e \qquad (4.2)$$

Now $i_e = i_b + i_c$ and Eqn 4.2 may therefore be written in the form

$$0 = i_c (R_l + r_c + r_e) + i_b r_e - \alpha r_c i_b - \alpha r_c i_c$$
$$= i_c [R_l + r_e + r_c (1 - \alpha)] + i_b (r_e - \alpha r_c) \qquad (4.3)$$

Current Gain

From Eqn 4.3 we can say

$$\beta = \frac{i_c}{i_b} = \frac{r_e - \alpha r_c}{R_l + r_e + r_c (1 - \alpha)}$$

Now r_e may be neglected in comparison with αr_c: normal values of R_l and r_e can be neglected in comparison with $r_c (1 - \alpha)$. We thus have

$$\beta = \frac{i_c}{i_b} = \frac{\alpha}{1 - \alpha}$$

$$\approx \frac{1}{1 - \alpha}$$

which agrees with the result derived on page 27. Values of β lie between 20 and 500 and tend to be higher for silicon than for germanium transistors. They fall off at low and high values of emitter current.

Input Resistance

As in the analysis of the common-base amplifier, we can obtain an expression for the input resistance of the amplifier by eliminating

i_c between Eqns 4.1 and 4.3 to obtain a relationship between i_b and v_{in}. From Eqn 4.3

$$i_c = -\frac{r_e - \alpha r_c}{R_l + r_e + (1-\alpha)r_c} \cdot i_b$$

Substituting for i_c in Eqn 4.1

$$v_{in} = i_b(R_s + r_b + r_e) - \frac{r_e(r_e - \alpha r_c)}{R_l + r_e + (1-\alpha)r_c} \cdot i_b$$

$$\therefore i_b = \frac{v_{in}}{R_s + r_b + r_e - \dfrac{r_e(r_e - \alpha r_c)}{R_l + r_e + (1-\alpha)r_c}}$$

Re-arranging

$$i_b = \frac{v_{in}}{R_s + r_b + r_e + \dfrac{r_e(\alpha r_c - r_e)}{R_l + r_e + (1-\alpha)r_c}} \tag{4.4}$$

If the input resistance of the amplifier is represented by r_i the input current i_b can be expressed

$$i_b = \frac{v_{in}}{R_s + r_i} \tag{4.5}$$

Comparing this with Eqn 4.4 we have

$$r_i = r_b + r_e + \frac{r_e(\alpha r_c - r_e)}{R_l + r_e + (1-\alpha)r_c} \tag{4.6}$$

which is a useful expression because it shows that r_i increases as R_l decreases. Eqn 4.6 can be simplified thus

$$r_i = r_b + \frac{r_e[R_l + r_e + (1-\alpha)r_c] + r_e(\alpha r_c - r_e)}{R_l + r_e + (1-\alpha)r_c}$$

$$= r_b + r_e \cdot \frac{R_l + r_c}{R_l + r_e + (1-\alpha)r_c}$$

The denominator can be simplified by neglecting r_e in comparison with the other terms which are much larger in practice. We then have the result

$$r_i = r_b + r_e \cdot \frac{R_l + r_c}{R_l + (1-\alpha)r_c} \tag{4.7}$$

Normally R_l is small compared with r_c and $(1-\alpha)r_c$. The input resistance is then given by

$$r_i = r_b + r_e \cdot \frac{r_c}{r_c(1-\alpha)} = r_b + \frac{r_e}{1-\alpha}$$

$$= r_b + \beta r_e \tag{4.8}$$

We have already shown (page 33) that r_e is inversely proportional to the mean emitter current I_e and we can thus say

$$r_i = r_b + \beta \cdot \frac{25 \, (\text{mV})}{I_e \, (\text{mA})}$$

For high values of β such as commonly encountered in silicon transistors the second term is large compared with the first. Thus

$$r_i \approx \beta \cdot \frac{25 \, (\text{mV})}{I_e \, (\text{mA})} \approx \frac{\beta}{g_m}$$

This is a useful result which shows that the input resistance is approximately inversely proportional to emitter current and directly proportional to β. Thus for a given emitter current the input resistance gives a direct measure of β for the transistor.

The above expression also shows that the input resistance of the common-emitter amplifier varies with emitter current (as in the common-base amplifier) and can cause distortion in a voltage amplifier if the source resistance is not large compared with the average value of input resistance.

When R_l is large compared with r_c and $(1-\alpha)r_c$ the input resistance has a value given by

$$r_i = r_b + r_e \cdot \frac{R_l}{R_l}$$

$$= r_b + r_e \tag{4.9}$$

Input Resistance of a Bipolar Transistor as Common-emitter Amplifier

To illustrate the range of values of input resistance likely to be encountered in practice in a common-emitter bipolar transistor amplifier, let us assume that $r_e = 25 \, \Omega$, $r_b = 300 \, \Omega$ and $\beta = 100$. Substituting in Eqn 4.8 to find the input resistance for short-circuited output terminals we have

$$r_i = 300 + 100 \times 25$$

$$= 2{,}800 \, \Omega$$

Substituting in Eqn 4.9 to obtain the input resistance for open-circuited output terminals we have

$$r_i = 300 + 25 \ \Omega$$

$$= 325 \ \Omega$$

These numerical examples show that the common-emitter amplifier has a higher input resistance than the common-base amplifier and that it decreases with increase in load resistance.

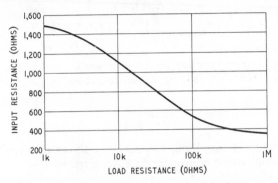

Fig. 4.2. Variation of input resistance with load resistance
for a common-emitter amplifier

The variation of input resistance with output collector load for bipolar transistors is illustrated in Fig. 4.2.

Output Resistance

We can obtain an expression for the output resistance of the common-emitter amplifier by eliminating i_b between Eqns 4.1 and 4.3 to obtain a relationship between v_{in} and i_c. From Eqn 4.3 we have

$$i_b = -\frac{R_l + r_e + r_c(1-\alpha)}{r_e - \alpha r_c} i_c$$

Substituting in Eqn 4.1

$$v_{in} = -\frac{R_l + r_e + r_c(1-\alpha)}{r_e - \alpha r_c}(R_s + r_b + r_e)i_c + i_c r_e$$

$$\therefore i_c = \frac{v_{in}}{-\dfrac{[R_l + r_e + r_c(1-\alpha)](R_s + r_b + r_e)}{r_e - \alpha r_c} + r_e}$$

$$= \frac{\dfrac{r_e - \alpha r_c}{R_s + r_b + r_e} \cdot v_{in}}{-[R_l + r_e + r_c(1-\alpha)] + \dfrac{r_e(r_e - \alpha r_c)}{R_s + r_b + r_e}}$$

This can be more conveniently written in the form

$$i_c = \frac{\dfrac{\alpha r_c - r_e}{R_s + r_b + r_e} \cdot v_{in}}{R_l + r_e + r_c(1-\alpha) + \dfrac{r_e(\alpha r_c - r_e)}{R_s + r_b + r_e}} \tag{4.10}$$

In a simple circuit containing a signal source of voltage v and of internal resistance r_o feeding a load resistance R_l the current i_c is given by

$$i_c = \frac{v}{R_l + r_o} \tag{4.11}$$

Comparison between Eqns 4.10 and 4.11 shows that the effective voltage acting in the equivalent circuit is given by

$$v = \frac{\alpha r_c - r_e}{R_s + r_b + r_e} \cdot v_{in}$$

and it is normally justifiable to assume that r_e is negligible compared with αr_c, that α can be taken as unity and that $(r_b + r_e)$ can be neglected in comparison with R_s. Making these assumptions we have

$$v \approx \frac{r_c}{R_s} \cdot v_{in}$$

and Eqn 4.11 now becomes

$$i_c = \frac{\dfrac{r_c}{R_s} \cdot v_{in}}{R_l + r_o}$$

both results identical to those for the common-base amplifier. But r_o for the common-emitter amplifier is approximately $1/\beta$ that for the common-base amplifier whereas R_s for the common-emitter amplifier must be approximately β times that for the common-base amplifier to limit distortion in a class-A amplifier to an acceptable level. These two effects cancel and the net result is that i_c has approximately the same value for a given value of v_{in} for common-emitter and common-base amplifiers.

If we compare Eqns 4.10 and 4.11 we obtain the following expression for the output resistance of the common-emitter amplifier:

$$r_o = r_e + r_c(1-\alpha) + \frac{r_e(\alpha r_c - r_e)}{R_s + r_b + r_e} \tag{4.12}$$

This expression shows that the output resistance depends on the source resistance (for a given transistor), decreasing as the source resistance is increased.

This expression can be simplified by combining the first term with the third; this gives the result

$$r_o = r_c(1-\alpha) + r_e \cdot \frac{R_s + r_b + \alpha r_c}{R_s + r_b + r_e} \tag{4.13}$$

To find an expression for the output resistance for short-circuited input terminals, let R_s approach zero in Eqn 4.13. We then have

$$r_o = r_c(1-\alpha) + r_e \cdot \frac{r_b + \alpha r_c}{r_b + r_e}$$

$$= r_c \cdot \frac{r_e + r_b(1-\alpha)}{r_e + r_b}$$

$$= r_c \cdot \frac{r_e + r_b/\beta}{r_e + r_b} \tag{4.14}$$

if we neglect the term $r_e r_b/(r_e + r_b)$ which is small compared with the others.

To find an expression for the output resistance for open-circuited input terminals let R_s approach infinity in Eqn 4.12. We then have

$$r_o = r_e + r_c(1-\alpha)$$

$$= r_e + r_c/\beta$$

and as r_e is small compared with r_c/β we can say

$$r_o = r_c/\beta \tag{4.15}$$

Output Resistance of a Bipolar Transistor as Common-emitter Amplifier

We can determine the range of values of output resistance for a common-emitter bipolar transistor amplifier by substituting the typical practical values $r_e = 25\ \Omega$, $r_b = 300\ \Omega$, $r_c = 1\ M\Omega$ and $\beta = 50$ in Eqns 4.14 and 4.15.

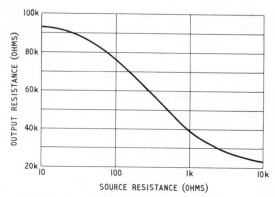

Fig. 4.3. Variation of output resistance with source resistance for a common-emitter amplifier

From Eqn 4.14 the output resistance for short-circuited input terminals is given by

$$r_o = 1,000,000 . \frac{25 + 300/50}{25 + 300} \,\Omega$$

$$= 95 \text{ k}\Omega$$

From Eqn 4.15 the output resistance for open-circuited input terminals is given by

$$r_o = \frac{1,000,000}{50} \,\Omega$$

$$= 20 \text{ k}\Omega$$

The variation in output resistance with source resistance for bipolar transistors is illustrated in Fig. 4.3.

Voltage Gain

From Fig. 4.1(b) we can see that the output voltage is given by $i_c R_l$. The voltage gain v_{out}/v_{in} is thus given by $i_c R_l/v_{in}$. From Eqn 4.10 i_c is given by

$$i_c = \frac{\dfrac{\alpha r_c - r_e}{R_s + r_b + r_e} . v_{in}}{R_l + r_e + r_c(1 - \alpha) + \dfrac{r_e(\alpha r_c - r_e)}{R_s + r_b + r_e}}$$

Hence

$$\frac{v_{out}}{v_{in}} = \frac{\dfrac{\alpha r_c - r_e}{R_s + r_b + r_e} \cdot R_l}{R_l + r_e + r_c(1-\alpha) + \dfrac{r_e(\alpha r_c - r_e)}{R_s + r_b + r_e}}$$

$$= \frac{(\alpha r_c - r_e)R_l}{[R_l + r_e + r_c(1-\alpha)](R_s + r_b + r_e) + r_e(\alpha r_c - r_e)} \tag{4.16}$$

Normally αr_c greatly exceeds r_e and this expression therefore gives a positive value for the voltage gain. But, to agree with the physics of the transistor, the collector current in Fig. 4.1(b) was assumed to be flowing in an anticlockwise direction whereas the base current was shown flowing in a clockwise direction. A positive value for the voltage gain thus implies that the output voltage is inverted with respect to the input voltage. Eqn 4.16 can be simplified by neglecting r_e in comparison with $R_l + r_c(1-\alpha)$ in numerator and denominator.

We then have

$$\frac{v_{out}}{v_{in}} = \frac{\alpha r_c R_l}{[R_l + r_c(1-\alpha)](R_s + r_b + r_e) + \alpha r_e r_c}$$

For values of R_l and r_e small compared with $r_c(1-\alpha)$ Eqn 4.16 may be written

$$\frac{v_{out}}{v_{in}} = \frac{\alpha r_c R_l}{r_c(1-\alpha)(R_s + r_b + r_e) + \alpha r_e r_c}$$

$$= \frac{\alpha R_l}{R_s(1-\alpha) + r_b(1-\alpha) + r_e}$$

If α in the numerator is taken as unity, this may be written

$$\frac{v_{out}}{v_{in}} = \frac{\beta R_l}{R_s + r_b + \beta r_e}$$

From Eqn 4.8 we can say

$$\frac{v_{out}}{v_{in}} = \frac{\beta R_l}{R_s + r_i} \tag{4.17}$$

which compares with Eqn 3.16 in the previous chapter for the common-base amplifier.

As pointed out in the previous chapter it is more useful to express voltage gain as the ratio of v_{out} to v_{be} and that an expression for this ratio can be obtained by putting $R_s = 0$. Thus we have

$$\frac{v_{out}}{v_{be}} = \beta \cdot \frac{R_l}{r_i}$$

and from page 56 $\beta/r_i = g_m$

$$\therefore \frac{v_{out}}{v_{be}} = g_m R_l$$

a useful result which compares directly with that for an f.e.t. and is used later. It should always be remembered, of course, that R_s must be large compared with r_i to keep distortion low even though R_s does not enter into the expression for v_{out}/v_{be}.

We have seen that a typical value for the input resistance of a transistor with $\beta = 100$ is 2·8 kΩ. Such a transistor if used for voltage amplification will require R_s to be at least 30 kΩ to minimise distortion due to input-resistance variation. A typical value for R_l is 5 kΩ and hence the voltage gain is given by

$$\frac{v_{out}}{v_{in}} = 100 \times \frac{5}{30}$$

$$= 17.$$

The voltage gain v_{out}/v_{be} is given by

$$\beta \cdot \frac{R_l}{r_i} = 100 \times \frac{5,000}{2,800}$$

$$= 180 \text{ approximately.}$$

Both values of gain are of the same order as those for the common-base amplifier.

Consider now the voltage gain of the common-emitter amplifier for load resistor values which are very large compared with $r_c(1-\alpha)$. If r_e is neglected in comparison with r_c, Eqn 4.16 becomes

$$\frac{v_{out}}{v_{in}} = \frac{\alpha r_c R_l}{R_l(R_s + r_b + r_e) + \alpha r_e r_c}$$

The second term in the denominator can be neglected in comparison with the first and α may be taken as unity giving

$$\frac{v_{out}}{v_{in}} = \frac{r_c}{R_s + r_b + r_e} \tag{4.18}$$

which is identical with Eqn 3.17 for the common-base amplifier.

Thus the voltage gain of a given transistor with a given large value of collector load resistance is approximately the same, no matter whether the transistor is connected up as a common-base or a common-emitter amplifier. The curve relating voltage gain with

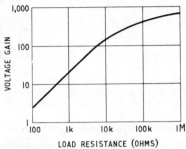

Fig. 4.4. Variation of voltage gain with load resistance for a common-emitter amplifier

load resistance is similar to that for the common-base amplifier and is given in Fig. 4.4.

Applications of Common-emitter Amplifiers

The common-emitter amplifier has a low input resistance, a high output resistance, high voltage gain and high current gain. It is the most widely used of the three fundamental bipolar transistor amplifiers and forms the basis of most types of amplifier, oscillator and pulse generator. Numerous examples of its applications are given in the remaining chapters of this book.

COMMON-SOURCE AMPLIFIERS

Amplifier Properties

The basic connections for the common-source amplifier are shown in Fig. 4.5(a). Fig. 4.5(b) shows the low-frequency equivalent circuit of the transistor with signal-source and load circuits.

The input resistance of the transistor, i.e. the resistance between gate and source terminals is clearly infinite: there is hence no possibility of feeding any current into the input terminals and we are concerned only with the performance of the transistor as a voltage amplifier.

The output resistance, i.e. the resistance between the drain and source terminals is simply r_{ds} the drain a.c. resistance, the constant-current generator being assumed to have infinite resistance.

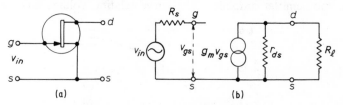

Fig. 4.5. Basic connections for a common-source amplifier (a) and the electrical equivalent with source and load circuit (b)

Because of the infinite input resistance there is no input current and hence no loss of input signal across R_s. Thus $v_{gs} = v_{in}$ and the current $g_m v_{in}$ divides between r_{ds} and R_l. The fraction i_{out} which enters R_l is given by

$$i_{out} = g_m v_{in} \cdot \frac{r_{ds}}{r_{ds} + R_l}$$

But $v_{out} = i_{out} R_l$

$$\therefore \frac{v_{out}}{v_{in}} = g_m \cdot \frac{r_{ds}}{r_{ds} + R_l}$$

Often R_l is small compared with r_{ds} and this can be simplified to

$$\frac{v_{out}}{v_{in}} = g_m R_l$$

Applications

The common-source amplifier has infinite input resistance and appreciable voltage gain. Both properties are useful and this is the most used of the f.e.t. linear circuits. Because of its low noise f.e.t.s are often used in low-level a.f. applications, e.g. in microphone head amplifiers where the high input resistance makes such amplifiers particularly suitable for following capacitor and piezo-electric microphones.

CHAPTER 5

Common-collector and Common-drain Amplifiers (Emitter and Source Followers)

COMMON COLLECTOR AMPLIFIERS

Introduction

The fundamental circuit of a common-collector amplifier is given in Fig. 5.1(a) in which R_s represents the resistance of the signal source and R_l represents the load resistance. For simplicity,

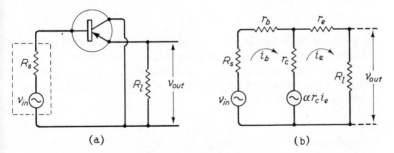

Fig. 5.1. The basic circuit for the common-collector transistor amplifier is given at (a), and the equivalent circuit at (b)

bias sources are omitted from this diagram. In Fig. 5.1(b) the transistor is represented by an equivalent network of resistances r_e, r_b and r_c together with a voltage generator of e.m.f. $\alpha r_c i_e$ where i_e

is the alternating current in the emitter circuit. If the base current i_b and emitter current i_e are both shown acting in clockwise directions as in Fig. 5.1(b), then the current in the common collector resistance r_c is the difference $(i_e - i_b)$. From Chapter 2 we know that this difference is the collector current i_c: all conventions are therefore satisfied in this diagram.

Current Gain

Applying Kirchhoff's laws to the circuit of Fig. 5.1(b) we have

$$v_{in} = i_b(r_b + r_c + R_s) + \alpha r_c i_e - i_e r_c$$
$$= i_b(r_b + r_c + R_s) - i_e r_c(1 - \alpha) \qquad (5.1)$$
$$0 = i_e(r_c + r_e + R_l) - \alpha r_c i_e - r_c i_b$$
$$= i_e[r_c(1 - \alpha) + r_e + R_l] - r_c i_b \qquad (5.2)$$

From Eqn 5.2 we have

$$\frac{i_e}{i_b} = \frac{r_c}{r_c(1 - \alpha) + r_e + R_l}$$

Normally r_e and R_l are both small compared with $r_c(1 - \alpha)$ and may be neglected in comparison with it. Thus the current gain of the common-collector circuit is given approximately by

$$\frac{i_e}{i_b} = \frac{1}{1 - \alpha}$$

As we have seen this is the approximation often used for β, the current gain of the common-emitter circuit. Thus the current gain of the common-collector and common-emitter circuit are very nearly equal: because i_c is only very slightly smaller than i_e this is not an unexpected result.

Input Resistance

We can obtain an expression for the input resistance if i_e is eliminated between Eqns 5.1 and 5.2 to obtain a relationship between v_{in} and i_b. From Eqn 5.2

$$i_e = i_b \cdot \frac{r_c}{r_c(1 - \alpha) + r_e + R_l}$$

Substituting for i_e in Eqn 5.1

$$v_{in} = i_b(r_b + r_c + R_s) - r_c(1-\alpha) \cdot \frac{i_b r_c}{r_c(1-\alpha) + r_e + R_l}$$

$$\therefore i_b = \frac{v_{in}}{r_b + r_c + R_s - \dfrac{r_c^2(1-\alpha)}{r_c(1-\alpha) + r_e + R_l}} \tag{5.3}$$

In a simple circuit containing a generator of resistance R_s and a load of resistance r_i the current is given by

$$i_b = \frac{v_{in}}{R_s + r_i} \tag{5.4}$$

Comparison of Eqns 5.3 and 5.4 gives the input resistance as

$$r_i = r_b + r_c - \frac{r_c^2(1-\alpha)}{r_c(1-\alpha) + r_e + R_l} \tag{5.5}$$

which shows that the input resistance of the common-collector

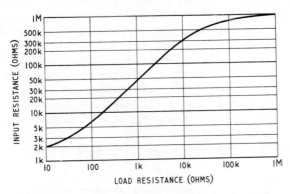

Fig. 5.2. Variation of input resistance with load resistance for a common-collector amplifier

amplifier depends on the parameters of the transistor (r_b, r_c, r_e and α) and on the load resistance R_l.

This expression can be simplified slightly by combining the second and third terms. We then have

$$r_i = r_b + \frac{r_c^2(1-\alpha) + r_e r_c + r_c R_l - r_c^2(1-\alpha)}{r_c(1-\alpha) + r_e + R_l}$$

$$= r_b + r_c \frac{r_e + R_l}{r_c(1-\alpha) + r_e + R_l} \qquad (5.6)$$

Consider the limiting value of input resistance as R_l approaches zero and in addition neglect r_e in comparison with $r_c(1-\alpha)$

$$r_i = r_b + \frac{r_e}{1-\alpha} \qquad (5.7)$$

$$= r_b + \beta r_e$$

which is, of course, the same result as for the common-emitter amplifier.

In practice values of R_l are likely to be large compared with r_e but small compared with r_c/β. If we neglect r_e in comparison with R_l in the numerator of Eqn 5.6 and R_l in comparison with $r_c(1-\alpha)$ in the denominator we have

$$r_i = r_b + \beta R_l \qquad (5.8)$$

$$\approx \beta R_l$$

a simple and useful result. Although it does not include r_e (as for the common-base and common-emitter amplifiers) the input resistance of the emitter follower is nevertheless dependent on emitter current because β normally varies with I_e. This does not mean however that the source resistance must be large compared with r_i when the emitter follower is used as a voltage amplifier. The circuit has a very large degree of negative feedback (sufficient, in fact, to reduce the voltage gain to less than unity) and this reduces distortion to a small value. The principal application of the emitter follower is as a voltage-operated device, fed from sources with a low output resistance.

If R_l is large compared with $r_c(1-\alpha)$ the input resistance is given by

$$r_i = r_b + r_c$$

but, of course, r_c is very large compared with r_b and thus we may say

$$r_i \approx r_c$$

Numerical Examples

Consider a transistor with $r_b = 300 \ \Omega$, $r_e = 25 \ \Omega$, $r_c = 1 \ M\Omega$ and $\beta = 50$. From Eqn 5.7 the input resistance for short-circuited output terminals is given by

$$r_i = 300 + 25 \times 50 \ \Omega$$

$$= 1,550 \ \Omega$$

which is the same value as for the common-emitter circuit. Suppose a load resistance of 1 kΩ is used in the emitter circuit. The input resistance is now given by Eqn 5.8

$$r_i = 50 \times 1{,}000 \ \Omega$$

$$= 50 \ \text{k}\Omega$$

For open-circuited output terminals the input resistance is equal to r_c, i.e. 1 MΩ.

These numerical examples show that the emitter follower can provide input resistances of 50 kΩ or more for load resistors of the order of 1 kΩ.

Output Resistance

We can obtain an expression for the output resistance of a common-collector transistor amplifier by eliminating i_b between Eqns 5.1 and 5.2, so as to obtain a relationship between i_e and v_{in}. From Eqn 5.2 we have

$$i_b = \frac{r_c(1-\alpha) + r_e + R_l}{r_c} \cdot i_e$$

Substituting for i_b in Eqn 5.1

$$v_{in} = \frac{(r_b + r_c + R_s)\left[r_c(1-\alpha) + r_e + R_l\right]}{r_c} i_e - i_e r_c(1-\alpha)$$

$$\therefore i_e = \frac{v_{in}}{\dfrac{(r_b + r_c + R_s)\left[r_c(1-\alpha) + r_e + R_l\right]}{r_c} - r_c(1-\alpha)}$$

Multiplying numerator and denominator by $r_c/(r_b + r_c + R_s)$ we have

$$i_e = \frac{\dfrac{r_c}{r_b + r_c + R_s} v_{in}}{R_l + r_e + r_c(1-\alpha) - \dfrac{r_c^2(1-\alpha)}{r_b + r_c + R_s}} \tag{5.9}$$

In a simple circuit containing a signal source of voltage v and internal resistance r_o feeding a load resistance R_l, the current i_e is given by

$$i_e = \frac{v}{R_l + r_o} \tag{5.10}$$

Comparison between Eqns 5.9 and 5.10 shows that the signal voltage effectively acting in the equivalent circuit is given by

$$v = \frac{r_c}{r_b + r_c + R_s} \cdot v_{in}$$

Normally r_c is large compared with $(r_b + R_s)$ and

$$v \approx v_{in}$$

Thus Eqn 5.10 becomes

$$i_e \approx \frac{v_{in}}{R_l + r_o}$$

and, from Eqn 5.13, $r_o \approx 1/g_m$. Thus we have the final important result that

$$i_e \approx \frac{v_{in}}{R_l + 1/g_m}$$

Thus the current in the load resistance R_l is the value which would flow if the signal input to the emitter follower acted directly on the load resistance and had a source resistance of $1/g_m$.

Comparing Eqns 5.9 with 5.10 we obtain the following expression for the output resistance r_o of the emitter follower

$$r_o = r_e + r_c(1-\alpha) - \frac{r_c^2(1-\alpha)}{r_b + r_c + R_s} \tag{5.11}$$

Combining the second and third terms

$$r_o = r_e + r_c(1-\alpha) \cdot \frac{r_b + R_s}{r_b + r_c + R_s} \tag{5.12}$$

When the source resistance is very small we have, neglecting r_b in comparison with r_c

$$r_o = r_e + r_b(1-\alpha)$$

It is often permissible to neglect $r_b(1-\alpha)$ in comparison with r_e. We then have

$$r_o \approx r_e$$

$$\approx \frac{1}{g_m} \tag{5.13}$$

Normal values of source resistance are large compared with r_b but small compared with r_c

$$\therefore r_o = r_e + R_s(1-\alpha)$$

$$= r_e + \frac{R_s}{\beta}$$

When R_s is large compared with r_c

$$r_o = r_e + r_c(1-\alpha)$$

$$= \frac{r_c}{\beta} \tag{5.14}$$

Some Numerical Examples

For a transistor with $r_b = 300\ \Omega$, $r_e = 25\ \Omega$, $r_c = 1\ M\Omega$ and $\beta = 50$ we have, using the above three simple approximations:

For very small source resistances

$$r_o = 25\ \Omega$$

For a source resistance of 1 kΩ

$$r_o = 25 + \frac{1,000}{50}\ \Omega$$

$$= 45\ \Omega$$

For a very high source resistance

$$r_o = \frac{1,000,000}{50}\ \Omega$$

$$= 20\ k\Omega$$

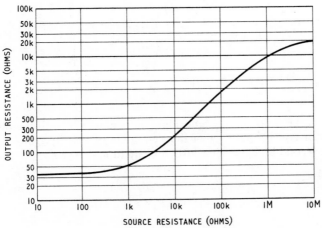

Fig. 5.3. Variation of output resistance with source resistance for a common-collector amplifier

The variation of output resistance with generator resistance is illustrated in Fig. 5.3. For small values of generator resistance, the output resistance is very low, being only slightly greater than the emitter resistance r_e. It is possible, for example, to have an output resistance of less than 50 Ω: a value as low as this is impossible from common-base or common-emitter transistor amplifiers.

Thus a common-collector amplifier with a low value of generator resistance and a high value of emitter load resistance can have a high value of input resistance and a low value of output resistance, conditions opposite to those normally encountered in transistor amplifiers and similar in fact to those which apply to a cathode follower using a thermionic valve.

Voltage Gain

From Fig. 5.1(b) we can see that the output voltage is given by $i_e R_l$. The voltage gain v_{out}/v_{in} is thus given by $i_e R_l/v_{in}$. From Eqn 5.9 i_e is given by

$$i_e = \frac{\dfrac{r_c}{r_b+r_c+R_s}v_{in}}{R_l+r_e+r_c(1-\alpha)-\dfrac{r_c^2(1-\alpha)}{r_b+r_c+R_s}}$$

Hence

$$\frac{v_{out}}{v_{in}} = \frac{\dfrac{r_c R_l}{r_b+r_c+R_s}}{R_l+r_e+r_c(1-\alpha)-\dfrac{r_c^2(1-\alpha)}{r_b+r_c+R_s}}$$

$$= \frac{r_c R_l}{[R_l+r_e+r_c(1-\alpha)](r_b+r_c+R_s)-r_c^2(1-\alpha)} \tag{5.15}$$

This can be simplified by ignoring r_e in comparison with the other terms in the first bracket of the denominator and by ignoring r_b in the second bracket. After further evaluation this gives

$$\frac{v_{out}}{v_{in}} = \frac{r_c R_l}{R_l r_c + R_l R_s + r_c(1-\alpha)R_s}$$

$$= \frac{R_l}{R_l + R_s/\beta + R_l R_s/r_c}$$

The third term in the denominator is normally negligible compared with the other two and thus we have

$$\frac{v_{out}}{v_{in}} = \frac{R_l}{R_l + R_s/\beta} \tag{5.16}$$

Thus the voltage gain is less than unity but R_s/β is small compared with R_l and the gain can usually be taken as unity with very little error.

Applications of Emitter Followers

The emitter follower has a high input resistance, a low output resistance, high current gain and unity voltage gain. Its main application is as a resistance converter, e.g. an emitter follower is often used as the first stage of a voltage amplifier to give a high input resistance and as the final stage to give a low output resistance. This may alternatively be regarded as a buffering action, e.g. the emitter follower first stage prevents the low input resistance of the second stage shunting the input signal source. Typical practical circuits for emitter followers are given in Chapter 7.

COMPARISON OF BIPOLAR TRANSISTOR AMPLIFIERS

To facilitate comparisons between the three fundamental types of transistor amplifier, the principal properties are summarised in Table 5.1. An approximate expression for each parameter is given in brackets.

Table 5.1

Fundamental properties of bipolar transistor amplifiers

	common-base	common-emitter	emitter follower
Input resistance	very low ($= r_e$)	low ($= \beta r_e$)	high ($= \beta R_l$)
Output resistance	very high ($= r_c$)	high ($= r_c/\beta$)	very low ($= 1/g_m$)
Current gain	unity ($= \alpha$)	high ($= \beta$)	high ($= \beta$)
Voltage gain	high	high	unity
Polarity of output signal relative to input signal	in phase	antiphase	in phase

COMMON-DRAIN AMPLIFIERS

Circuit Properties

The basic circuit for a common-drain amplifier (source follower) is given in Fig. 5.4(a) and its electrical equivalent at (b). Fig. 5.4(a)

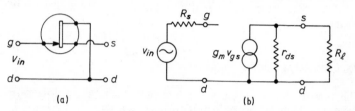

Fig. 5.4. *Basic connections for a source follower amplifier (a) and the electrical equivalent with source and load circuit (b)*

shows that the input resistance is infinite: the output resistance can be calculated in the following way.

As in the common-source circuit the current $g_m v_{gs}$ splits between r_{ds} and R_l the fraction (i_{out}) entering R_l being given by

$$i_{out} = g_m v_{gs} \cdot \frac{r_{ds}}{r_{ds} + R_l}$$

Normally R_l is small compared with r_{ds} and thus we can say

$$i_{out} = g_m v_{gs} \qquad (5.17)$$

Now $v_{out} = i_{out} R_l$

$$\therefore \frac{v_{out}}{v_{gs}} = g_m R_l$$

This is, of course, the voltage gain of the common-source amplifier. Adding unity to both sides we have

$$\frac{v_{out} + v_{gs}}{v_{gs}} = 1 + g_m R_l$$

Now, from Fig. 5.4(a), $(v_{out} + v_{gs}) = v_{in}$

$$\therefore v_{in} = v_{gs}(1 + g_m R_l) \qquad (5.18)$$

Substituting for v_{gs} in Eqn 5.17

$$i_{out} = g_m \cdot \frac{v_{in}}{1 + g_m R_l}$$

$$= \frac{v_{in}}{R_l + 1/g_m}$$

which shows that the load is effectively fed from a generator of voltage v_{in} and output resistance $1/g_m$. The source follower thus has unity voltage gain and an output resistance of $1/g_m$. The voltage gain can alternatively be calculated thus. From Fig. 5.4(a)

$$v_{in} = v_{gs} + v_{out}$$

Substituting for v_{gs} from Eqn 5.18

$$v_{in} = \frac{v_{in}}{1 + g_m R_l} + v_{out}$$

from which

$$\frac{v_{out}}{v_{in}} = \frac{g_m R_l}{1 + g_m R_l}$$

Now $g_m R_l$ is the voltage gain of the common-source stage and is normally large compared with unity. Thus the voltage gain of the source follower is very nearly unity.

Applications

The source follower has an infinite input resistance, a low output resistance and a voltage gain of approximately unity. Its only advantage over the common-source amplifier is the low output resistance and it is used wherever such a property is essential; otherwise the common-source amplifier is preferred.

COMPARISON OF FIELD-EFFECT TRANSISTOR AMPLIFIERS

To facilitate comparisons between the three fundamental types of f.e.t. amplifier the principal properties are summarised in Table 5.2. An approximate expression for each parameter is given in brackets.

Table 5.2

Fundamental properties of f.e.t. amplifiers

	common-gate	common-source	source follower
Input resistance	very low $(= 1/g_m)$	infinite	infinite
Output resistance	high $(= r_{ds})$	high $(= r_{ds})$	very low $(= 1/g_m)$
Current gain	unity	—	—
Voltage gain	fair $(= g_m R_l)$	fair $(= g_m R_l)$	unity
Polarity of output signal relative to input signal	in phase	antiphase	in phase

CHAPTER 6

Bias and D.C. Stabilisation

BIPOLAR TRANSISTORS

Introduction

It is usual to begin the design of a bipolar transistor amplifying stage by choosing a value of mean collector current which enables the required current swing and/or voltage swing to be delivered with an acceptable degree of linearity. The next step is to devise a biasing circuit to give this particular value of mean collector current.

The most obvious way of biasing a bipolar transistor is by the

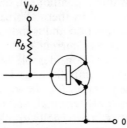

Fig. 6.1. Simple bias circuit which fixes base current

simple circuit of Fig. 6.1 in which a resistor R_b is connected between the base of the transistor and source of steady voltage V_{bb}. Certainly it is possible by adjustment of R_b and/or V_{bb} to set the mean collector current at the desired value but such adjustments to individual transistors would be tedious if the amplifiers are to be produced in any quantity. The fundamental disadvantage of the circuit of Fig. 6.1 is that it fixes the *base current** not the *collector current*.

---

* Provided V_{bb} is large compared with V_{be}, and R_b is large compared with the transistor input resistance the base current is given approximately by V_{bb}/R_b.

The collector current is β times the base current and values of β for a given type of transistor can differ by $\pm 50\%$ from the nominal value. Thus if the nominal β is 100, values as low as 50 and as high as 150 are possible and, for a given base current, the collector current can have any value within a range of 3 : 1. Smaller variations in β also occur as a result of temperature changes. Moreover in certain circuits, particularly those using power transistors, the collector current can increase significantly with temperature as a result of variations in the value of V_{be} which, for all bipolar transistors, increases at the rate of 2·5 mV per °C. It is these variations in β and V_{be} which make the circuit of Fig. 6.1 unsuitable for use in mass-produced equipment. What is wanted for this purpose is a biasing circuit which, without using preset components:

(a) gives the desired value of mean collector current in spite of manufacturing spreads in transistor parameters and

(b) maintains this mean current in spite of variations in parameters due to temperature changes.

Circuits which fulfil these two purposes are said to provide *d.c. stabilisation* of the operating point and this chapter describes the most commonly-used circuits.

Leakage Current

Unfortunately the amplified input current is not the only component of the collector current of a bipolar transistor. There is also an unwanted component, known as the *leakage current*, which is generated by thermal breakdown of covalent bonds in the collector-base junction and is independent of the input current.

Leakage current increases rapidly with rise in temperature and can be of considerable importance in common-emitter amplifiers. In such amplifiers the leakage current generated in the collector-base junction I_{CBO} flows through the base-emitter junction to reach the supply terminals and in so doing is amplified β times in the same way as an externally-applied input current. Thus the leakage current in a common-emitter stage can be considerable: for a small germanium transistor it can rise from 250 μA at 25°C to 2·5 mA at 55°C. Such values can exceed the useful current and seriously limit the performance of the transistor by preventing the full excursion of collector current and collector voltage. Moreover the leakage current heats the collector-base junction thus accelerating the breakdown of covalent bonds and still further increasing leakage current. Thus a regenerative build-up of temperature and leakage current (known as *thermal runaway*) can occur and this, unless checked (e.g.

by a stabilising circuit) can damage or even destroy the transistor. One of the most serious disadvantages of the circuit of Fig. 6.1 is that it gives no protection against thermal runaway.

Leakage current is negligible in silicon transistors at normal temperatures but d.c. stabilisation is still necessary to minimise the effects of the spreads in transistor parameters and the temperature dependence of β and V_{be}.

Stability Factor

The effectiveness of circuits for stabilising mean collector current can be expressed by a stability factor K which may be defined in the following way:

$$K = \frac{\text{collector current in stabilised circuit}}{\text{collector current in unstabilised circuit}}$$

K is thus unity for an unstabilised circuit and is less than unity for stabilised circuits, the smallness of K being a measure of the success of the circuit in limiting increases in mean current.

Basic Circuits for d.c. Stabilisation

Most stabilising circuits achieve their object by d.c. negative feedback: the collector or emitter current is used to generate a signal which is returned to the base so as to oppose any changes in the mean value of the collector or emitter current. In practice the

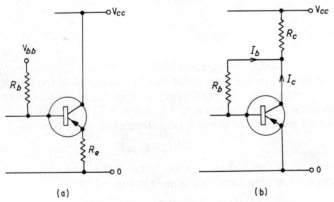

(a)   (b)

Fig. 6.2. The two basic circuits for d.c. stabilisation

transistor current is passed through a resistor and the voltage generated across the resistor is used as a source of base bias current. The resistor can be in the emitter circuit as shown in Fig. 6.2(a) or in the collector circuit as in Fig. 6.2(b). The way in which these circuits achieve stabilisation is perhaps more easily understood from Fig. 6.2(b): any increase in collector current causes an increased voltage across R_c, a decrease in the collector-emitter voltage and thus a smaller base current which opposes the initial rise in collector current. Provided R_c equals R_e the two circuits of Fig. 6.2 have

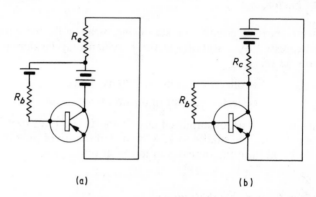

(a)   (b)

Fig. 6.3. The two basic circuits redrawn to show their similarity

identical performances: this can be shown by redrawing the two circuits in the form given in Fig. 6.3 from which it is clear that if the d.c. resistance of the batteries is neglected the two circuits are identical.

Calculation of stability factor

Provided R_c equals R_e therefore both circuits have the same stabilising effect and the stability factor can be calculated in the following way in which the equations apply to Fig. 6.2(b). The calculation applies equally to the circuit of Fig. 6.2(a) if $R_c = R_e$ and $V_{bb} = V_{cc}$.

For an unstabilised common-emitter amplifier we have

$$I_c = \beta(I_b + I_{CBO}) \tag{6.1}$$

where I_{CBO} is the leakage current generated in the collector-base junction. Summing steady potentials in Fig. 6.2(b)

$$R_c(I_c + I_b) + R_b I_b + V_{be} = V_{cc} \tag{6.2}$$

where V_{cc} is the supply voltage and V_{be} is the base-emitter voltage of the transistor.

From Eqn 6.1 we have

$$I_b = \frac{I_c}{\beta} - I_{CBO}$$

Substituting for I_b in Eqn 6.2

$$I_c[(\beta+1)R_c + R_b] = I_{CBO}\beta(R_c + R_b) + \beta(V_{cc} - V_{be})$$

Rearranging

$$I_c = \frac{\beta I_{CBO}}{1 + \beta R_c/(R_b + R_c)} + \frac{\beta(V_{cc} - V_{be})/R_b}{1 + (\beta+1)R_c/R_b} \tag{6.3}$$

The first term in expression 6.3 gives the leakage current in the stabilised circuit and the second term gives the mean value of the useful component of the collector current. The first term is of interest in the design of germanium transistor circuits and the second is of concern mainly in the stability considerations of silicon transistor circuits.

Stability factor for leakage current

From Eqn 6.3 we have

$$\text{leakage current} = \frac{\beta I_{CBO}}{1 + \beta R_c/(R_b + R_c)}$$

βI_{CBO} is, of course, the leakage current of the unstabilised common-emitter amplifier

$$\therefore K = \frac{1}{1 + \beta R_c/(R_b + R_c)} \tag{6.4}$$

If R_c is very small or if R_b is very large, the feedback providing stabilisation disappears and K is 1 as would be expected for an unstabilised common-emitter circuit. On the other hand if R_b is very small or if R_c is very large, K approaches $1/\beta$ which represents the best stability factor achievable with this circuit.

Stability factor for useful component of collector current

The second term in expression 6.3 shows how the mean value of the useful component of collector current increases with increase in β.

If R_c is very small or if R_b is very large (to eliminate feedback) we have

$$\text{mean value of useful component of } I_c = \frac{\beta(V_{cc} - V_{be})}{R_b} \qquad (6.5)$$

showing that this component is directly proportional to β as would be expected in an unstabilised circuit. On the other hand if R_b is very small or R_c is very large we have

$$\text{mean value of useful component of } I_c = \frac{V_{cc} - V_{be}}{R_c}$$

i.e. the current is independent of β and stabilisation is perfect. Expression 6.5 is the numerator of the second term in expression 6.3 and thus we have that the stabilisation factor is given by

$$K = \frac{1}{1 + (\beta + 1)R_c/R_b} \qquad (6.6)$$

which is similar to Eqn 6.4 for leakage current: in fact the two expressions are approximately equal if, as in many circuits, R_c is small compared with R_b.

Estimation of stability factor from circuit diagram

Eqns 6.4 and 6.6 are both of the form

$$K = \frac{1}{1 + \beta F} \qquad (6.7)$$

where F is the fraction of the collector current which is fed back to the base. In Fig. 6.2(b) the collector current I_c splits at the junction of R_c and R_b. The current entering R_b is given by

$$I_c \cdot \frac{R_c}{R_b + R_c}$$

and this, of course, is the base current I_b. Thus the fraction of I_c used as feedback is $R_c/(R_b + R_c)$ which agrees with expression 6.3. This is a useful method of deriving stability factors which often enables the factor to be calculated directly from the component values on a circuit diagram. The circuit diagram should be reduced to the form shown in Fig. 6.3 which shows the split of collector current and enables F to be computed from the ratio $R_c/(R_c + R_b)$. The stability factor is then given by Eqn 6.7.

Minimising signal-frequency feedback

In Fig. 6.2 the feedback circuit cannot distinguish between direct current and signal-frequency current. Thus the gain of wanted signals

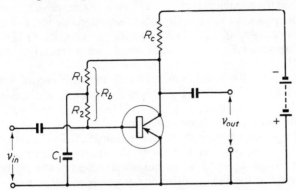

Fig. 6.4. To avoid signal-frequency negative feedback due to R_b, decoupling may be introduced as shown here

is reduced in the same ratio as the leakage current (or as the increase in mean current due to increase in β). If this reduction in signal-frequency gain is not wanted it may be minimised by decoupling. If the signal-frequency output is taken from the collector circuit

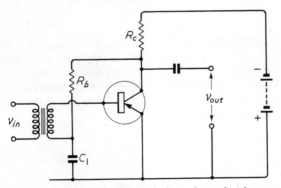

Fig. 6.5. Method of decoupling which can be used with trans-former coupling

in Fig. 6.2(a) R_e can be shunted by a low-reactance capacitor to minimise signal-frequency feedback. If the signal-frequency output is taken from the collector circuit in Fig. 6.2(b) it is not practical to decouple R_c and a more usual method is to construct R_b of two resistors R_1 and R_2 in series, the junction being decoupled to emitter as shown in Fig. 6.4. If R_1 is small it reduces the effective collector load resistance of the amplifier and if R_2 is small it lowers the input resistance. Usually therefore R_1 and R_2 are made approximately equal and the capacitor is chosen to have a reactance small compared with the resistance value at the lowest signal frequency. For example if R_b has a value of 100 kΩ R_1 and R_2 can each be 50 kΩ and for an a.f. amplifier C_1 can be 2 μF which has a reactance of 1·6 kΩ at 50 Hz.

If the transistor is fed from a transformer one end of R_b can be decoupled as shown in Fig. 6.5. Signal-frequency feedback is mini-mised by the decoupling capacitor C_1 but in this circuit R_b has no shunting effect on the input resistance.

If a large output voltage swing is required the quiescent collector potential should be made one-half the supply voltage. Thus if in an npn circuit $V_{cc} = 12$ V the quiescent potential should be made 6 V. Suppose the mean collector current is required to be 3 mA. We have immediately that R_c is $6/(3 \times 10^{-3}) = 2$ kΩ. If a germanium transistor is used V_{be} may be taken as zero and the voltage across R_b is also 6. Suppose β is 100. I_b is then 3 mA/100, i.e. 30 μA and R_b is given by $6/(30 \times 10^{-6}) = 200$ kΩ. The stability factor for this circuit is, from Eqn 6.6, given by

$$K = \frac{1}{1 + 100 \times 2/(200 + 2)}$$

$$= 0·5$$

Without stabilisation $K = 1$ and thus this circuit has succeeded only in halving the leakage current—a very poor performance.

The stabilising circuit of Fig. 6.2(a) is the more used because signal-frequency decoupling is simple and because the collector is free for signal-frequency components. Most d.c. stabilising circuits are based on Fig. 6.2(a).

Use of Tapped Battery

We have seen that maximum d.c. stability is obtained from the circuit of Fig. 6.2(a) by making $R_b = 0$. This means that the transistor base is connected directly to the base bias supply which can be a tapping on the collector supply battery. A practical example of such a circuit is given in Fig. 6.6.

To estimate the component values needed and stability factor obtainable suppose the transistor is a silicon type and is required to have a mean collector current of 5 mA. Suppose also that a 3-volt tapping on the battery is chosen for base bias. The base-emitter voltage is 0·7 and the voltage across R_e is 2·3 giving R_e as $2·3/(5 \times 10^{-3})$ = 460 ohms. R_b is effectively the d.c. resistance of the transformer secondary winding and a typical value is 100 ohms. From Eqn 6.6 we can calculate the stability factor as follows:

$$K = \frac{1}{1 + 100 \times 460/(460 + 100)}$$

$$= 0·012$$

a reduction of variations in collector current to less than one-eightieth of their unstabilised value, a very satisfactory performance which is not far short of the maximum obtainable ($K = 0·01$).

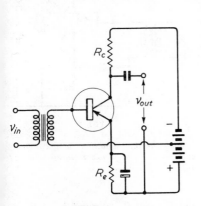

Fig. 6.6. *Tapped-battery method of d.c. stabilisation in a transformer-coupled amplifier*

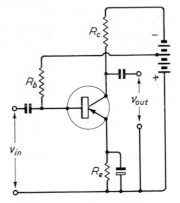

Fig. 6.7. *Tapped-battery method of d.c. stabilisation in an RC-coupled amplifier*

Even if the transformer is replaced by a base bias resistor (as shown in Fig. 6.7) a worthwhile reduction in collector current variations can be achieved. R_b can be as low as 3·3 kΩ without too drastic a reduction in the input resistance of the amplifier. For the same values of β and R_e as before

$$K = \frac{1}{1 + 100 \times 460/(460 + 3,300)}$$

$$= 0·076$$

a reduction of collector current variations to less than one-tenth of their unstabilised value.

Potential Divider and Emitter Resistor Circuit

It is not always possible or convenient to tap the supply battery to obtain base bias but an alternative source of steady potential is a potential divider $R_1 R_2$ connected across the collector supply as shown in Fig. 6.8.

Two basic types of this circuit exist. In Fig. 6.8(a) the circuit is arranged for RC coupling from the previous stage: a feature of this

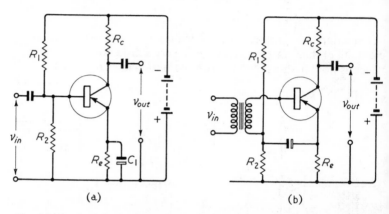

Fig. 6.8. *Potential-divider method of stabilising the d.c. conditions in a common-emitter amplifier (a) in an RC-coupled amplifier and (b) in a transformer-coupled amplifier*

arrangement is that the resistance of R_1 and R_2 in parallel is effectively shunted across the input circuit of the transistor. This parallel resistance should not, therefore, be too small. In Fig. 6.8(b) the circuit is arranged for transformer coupling from the previous stage: the parallel resistance of R_1 and R_2 does not now enter into input-resistance considerations.

If these circuits are redrawn in the form shown in Fig. 6.9 we can see immediately that they are of the same type as Figs. 6.2 and 6.3 in which R_b' is made up of R_1 and R_2 in parallel. Thus we can at once estimate the stability factor. The output current division ratio F (Eqn 6.7) is $R_e/(R_b' + R_e)$ and hence

$$K = \frac{1}{1 + \beta R_e/(R_b' + R_e)}$$

$$\text{where } R_b' = \frac{R_1 R_2}{R_1 + R_2}$$

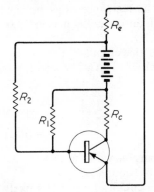

Fig. 6.9. The circuit of Fig. 6.8. redrawn to illustrate the feedback due to R_e

Good stability thus requires low values of R_1 and R_2 but this means a heavy drain on the supply. In practice a compromise solution is adopted as indicated in the following numerical example.

Design of a potential-divider circuit

Consider first the design of such a circuit for a germanium transistor. Because V_{be} is very small it is possible to work with a voltage as low as 1 V across R_2 and to assume that the same voltage is generated across R_e.

Let us assume R_e to be 1 kΩ: this is a convenient value because, for an emitter current of 1 mA, only 1 V of the collector supply voltage is lost. This leaves in a typical circuit with a 6 V supply, 5 V for the transistor and its load resistor. If $\beta = 50$ the base current is 1/50 mA, i.e. 20 μA. This flows through R_1 in addition to the bleed current which flows through R_1 and R_2 from the collector supply. For good d.c. stability the potential at the junction of R_1 and R_2 must be steady in spite of variations in base current and this is achieved by making the parallel resistance of R_1 and R_2 small: this implies that the bleed current must be large compared with the base current. The bleed current can therefore be 200 μA which is ten times the base current but is only one-fifth the collector current. Thus the total current in R_1 is 220 μA. The voltage across R_1 is 5 V because the potential at the junction of R_1 and R_2 does not differ appreciably from the emitter potential. Hence R_1 is given by $5/(220 \times 10^{-6}) = 23$ kΩ approximately.

The voltage drop across R_2 is 1 V and the current in it is 200 μA, giving the value of R_2 as 5 kΩ. C_1 should have a reactance small enough to avoid negative feedback and consequent fall in signal-frequency gain even at the lowest operating frequency. To achieve this the reactance must be small compared with r_e, say 25 Ω. In an a.f. amplifier C_1 may be 500 μF which has a reactance of 6·5 Ω at 50 Hz.

For this circuit R'_b is given by $R_1 R_2/(R_1 + R_2)$, i.e. $5,000 \times 23,000/28,000 = 4·1$ kΩ. The stability factor is given by Eqn 6.6:

$$K = \frac{1}{1 + 50 \times 1/(4·1 + 1)}$$

$$= 0·094$$

comparable with the stability achieved by the tapped battery circuit with RC coupling.

The circuit of Fig. 6.8 is likely to be used with silicon transistors to minimise the effects of the spread in β. Allowance must now be made for the higher offset voltage of silicon transistors. For satisfactory operation of the circuit the voltage across R_e must be large compared with V_{be} and this necessitates a voltage of at least 3 V across R_e: this is approximately half the supply voltage commonly used in battery-operated receivers. This limitation is not so serious in mains-driven equipment where higher supply voltages are easier to obtain. For example if 24 V is available 7 V could be allowed across R_e. Let the required mean collector current be 4 mA. R_e is then $7/(4 \times 10^{-3})$, approximately 1·8 kΩ. If β is 150, I_b is 4/150 mA, that is 27 μA. Let the potential divider take 0·5 mA from the supply. I_b can be neglected in comparison with this and we can say that $(R_1 + R_2)$ is $24/(0·5 \times 10^{-3})$, i.e. 48 kΩ. The voltage required across R_2 is 7·7 (the offset voltage being taken as 0·7) and thus R_2 is given by $7·7/(0·5 \times 10^{-3})$, i.e. 15·4 kΩ. R_1 is therefore $48 - 15·4 = 32·6$ kΩ.

R'_b for this circuit is the parallel resistance of R_1 and R_2, i.e. $15·4 \times 32·6/48 = 10·5$ kΩ. From Eqn 6.6 the stability factor is given by

$$K = \frac{1}{1 + (150 + 1)2/10·5}$$

$$= 0·034$$

showing a thirty-fold reduction in the effects of a change of β. Thus a $\pm 50\%$ spread in β is reduced by this circuit to $\pm 1·6\%$.

D.C. Stability in Two-stage Amplifiers

It is clear from Fig. 6.9 that the performance of the potential-divider circuit could be improved by returning R_1 to the collector: by so doing we ensure that R_c makes a contribution to the d.c. feedback in addition to that provided by R_e. The circuit so produced is shown in Fig. 6.10: in Fig. 6.11 it is redrawn to make the feedback paths more obvious. The improvement in stability obtained in this circuit can be estimated in the following way.

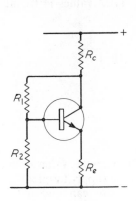

Fig. 6.10. Potential divider returned to collector

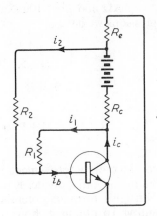

Fig. 6.11. The circuit of Fig. 6.10 redrawn to show feedback paths

The current I_c leaving the collector in Fig. 6.11 splits between R_c and R_1. The current I_1 entering R_1 is given approximately by

$$I_1 = I_c \cdot \frac{R_e + R_c}{R_e + R_c + R_1}$$

in which, for simplicity, the shunting effect of R_2 on R_e is neglected. The current in R_c is now

$$I_c \cdot \frac{R_1}{R_e + R_c + R_1}$$

and this splits at the junction of R_e and R_2. The current I_2 in R_2 is given by

$$I_2 = I_c \cdot \frac{R_1}{R_e + R_c + R_1} \cdot \frac{R_e}{R_e + R_2}$$

I_b is the sum of I_1 and I_2. Thus the feedback fraction F is given by

$$F = \frac{I_b}{I_c} = \frac{R_e + R_c}{R_e + R_c + R_1} + \frac{R_1}{R_e + R_c + R_1} \cdot \frac{R_e}{R_e + R_2}$$

and the stability factor is, from Eqn 6.7:

$$K = \frac{1}{1 + \beta F}$$

Typical component values are $R_e = 1 \text{ k}\Omega$, $R_c = 5 \text{ k}\Omega$, $R_1 = 10 \text{ k}\Omega$, $R_2 = 5 \text{ k}\Omega$ and $\beta = 100$. Substituting these values in the above expressions we have

$$F = \frac{1+5}{1+5+10} + \frac{10}{1+5+10} \cdot \frac{1}{1+5}$$

$$= 0.48$$

$$K = \frac{1}{1 + 100 \times 0.48}$$

$$= 0.02$$

a considerable improvement in stability over the value obtained when the potential divider is returned to the supply terminal.

Thus this form of potential-divider circuit can give excellent stability. To obtain it, however, the shunting effect of R_1 on R_c must be tolerated. Moreover if signal-frequency feedback via R_1 is to be avoided R_1 must be constructed of two resistors in series with their junction decoupled as shown in Fig. 6.4. Both disadvantages can be overcome by the use of an emitter follower as shown in Fig. 6.12. The high input resistance of TR2 minimises the shunting effect on R_c and the low output resistance makes possible low values of R_1 and R_2. If the emitter circuit of TR2 is decoupled at signal

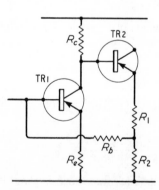

Fig. 6.12. Potential-divider circuit using an emitter follower

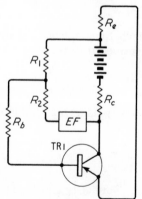

Fig. 6.13. Fig. 6.12 redrawn to facilitate calculation of stability factor

frequencies, this transistor can be used as a common-emitter amplifier and there is no need to include an additional transistor in the
circuit purely to give good d.c. stabilisation. If, however, $(R_1 + R_2)$
is decoupled it is necessary to introduce a series resistor R_b to avoid
decoupling TR1 base as shown in Fig. 6.12. Some examples of this
circuit are given in the next chapter.

The stability factor obtainable from the circuit of Fig. 6.12 can
be calculated from the redrawn version of the circuit shown in
Fig. 6.13.

There is effectively no split of collector current at the junction of
R_c and the emitter follower because of the high resistance of the
emitter follower. It is best to calculate I_b from a knowledge of the
voltage applied to R_b. From the emitter follower one component of
this voltage is $R_1/(R_1 + R_2)$ of that generated across R_c by I_c: this
we can call rI_cR_c. The second component of the voltage is that
generated across the parallel resistance of R_e and R_b by I_c: it is
assumed that R_1 is negligibly small. Thus we have

$$I_b = \frac{rR_cI_c + \dfrac{R_eR_b}{R_e + R_b} \cdot I_c}{R_b}$$

$$= I_c \left(\frac{rR_c}{R_b} + \frac{R_e}{R_e + R_b} \right)$$

giving

$$F = \frac{rR_c}{R_b} + \frac{R_e}{R_e + R_b}$$

Thus the stability factor is given by

$$K = \frac{1}{1 + \beta \left(\dfrac{rR_c}{R_b} + \dfrac{R_e}{R_e + R_b} \right)}$$

Typical practical values are $r = 0.5$, $R_c = 5$ kΩ, $R_e = 5$ kΩ,
$R_b = 3.3$ kΩ and $\beta = 100$. Substituting in the above expression

$$K = \frac{1}{1 + 100 \left(\dfrac{0.5 \times 5}{3.3} + \dfrac{5}{5 + 3.3} \right)}$$

$$= 0.0073$$

This represents exceptionally good stability, variations in mean
collector current in TR1 being reduced to less than one-hundredth

of their unstabilised value. Transistors are often used in cascade in amplifiers and it is common practice to adopt the circuit of Fig. 6.12 to achieve high d.c. stability.

Use of Diodes to Compensate for Falling Battery Voltage

The stabilising circuits described above are useful in reducing leakage current and the effects of spreads in β but they do nothing towards making the mean collector current independent of the supply voltage. Such circuits are useful therefore when a stabilised supply is available but in battery-operated equipment some means is required of compensating for the effects of falling battery voltage. One method which uses the circuit of Fig. 6.8 is to arrange for the resistance of R_2 to increase as the current through it falls, e.g. by using a non-linear device in place of R_2 (or in parallel with it): a suitable device is a forward-biased semiconductor diode. Such diodes are used in battery-operated portable receivers to stabilise the quiescent current of class-B output stages.

Use of Diodes for Temperature Compensation

Diodes connected in the base circuits of transistors are also used to stabilise the collector current against temperature changes. Very good stabilisation can be achieved by using in the compensating circuit a diode of the same material as the transistors to be stabilised. As temperature changes the voltage across the diode changes by the right amount to keep the current constant in the transistors. If the transistors are mounted on a heat sink the diodes should preferably be mounted close to them on the same heat sink.

Diodes are extensively used in integrated circuits for stabilising collector currents and here, of course, they share a silicon chip with the transistors to be stabilised. They are thus in intimate thermal contact with the circuit to be monitored and are capable of maintaining satisfactory performance over a temperature range as wide as $-55°C$ to $125°C$.

Use of Temperature-dependent Resistor for Temperature Compensation

As an alternative to using a diode in the base circuit it is possible to use a resistor with a positive temperature coefficient in the emitter circuit of the transistor to be stabilised. The technique is often used with power transistors as a means of stabilising collector currents of the order of 0·5 A. This is large compared with the leakage current and variations in collector current are almost entirely due to changes in V_{be} with temperature. The mean value of the useful component can be kept constant by adjustment of the base-emitter voltage and a change of approximately 2·5 mV per °C is required by germanium and silicon transistors. Thus the stabilising circuits for power transistors should be designed to apply a correction of this value to the base-emitter voltage.

A simple method of effecting this compensation is to use an external emitter resistor of pure metal. Such resistors have a positive temperature coefficient and a rise in temperature causes the external emitter resistance to increase, thus increasing the voltage across this resistance. This in turn reduces the base-emitter voltage and, if the base potential is constant, tends to maintain the collector current constant.

The temperature coefficient of electrical resistance of copper is approximately 0·004 per °C: if the emitter current is assumed constant, the voltage across a copper emitter resistor therefore increases by 0·004 of its initial value per °C. If the initial voltage is unity the change in emitter voltage is 4 mV per °C. To offset a 2·5 mV change in base-emitter voltage, an initial emitter voltage of 2·5/4, i.e. approximately 0·6 V is needed. If the mean emitter current is 0·5 A (as is likely in a transistor with 5 W dissipation) the emitter resistance should be 0·6/0·5, i.e. 1·2 Ω a convenient value to construct of copper wire. The fixed base potential is usually achieved by the use of a resistive potential divider.

FIELD-EFFECT TRANSISTORS

Introduction

It is usual to begin the design of an amplifying circuit using an f.e.t. by choosing a value of mean drain current which will enable the required output current swing and/or output voltage swing to be achieved with an acceptable degree of linearity. The next step is to devise a biasing circuit to give the required value of mean drain current.

The obvious way to bias an f.e.t. is by the simple circuit of Fig. 6.14 which contains a resistor R_g connected between the gate and a source of constant voltage V_{gg}. The gate current in an f.e.t. is very small indeed and R_g can be very high, e.g. 100 MΩ, without significantly affecting the gate voltage. As pointed out in Chapter 2 the bias voltage for an enhancement f.e.t. lies between the source and drain voltages whereas for a depletion f.e.t. it lies outside the range of the drain-source voltage. Typical figures for both types of f.e.t. are given in Fig. 6.14.

Certainly by adjusting V_{gg} it is possible to set the mean drain current at the desired value but such adjustments to individual

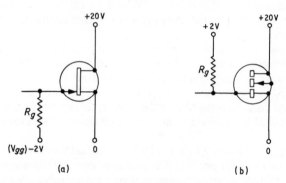

(a)   (b)

Fig. 6.14. Simple bias circuit for fixing gate-source voltage
(a) in depletion and (b) in enhancement f.e.t.s

transistors would be tedious if the circuits are to be produced in any quantity. The fundamental disadvantage of this simple biasing circuit is that it fixes the *gate voltage* not the *drain current*. The drain current for a given type of f.e.t. and a given gate-source voltage can have any value within a range of 3:1 due to manufacturing spreads in transistor parameters and to changes in parameters with temperature. In fact drain current may increase or decrease as temperature changes but there is no danger of thermal runaway as in bipolar transistors. Ideally what is wanted is a biasing circuit which will enable a desired value of mean drain current to be obtained in spite of manufacturing spreads and which will also maintain this mean current in spite of changes in parameters.

Such d.c. stabilisation can be achieved, as for bipolar transistors, by d.c. negative feedback, e.g. by passing the drain current through a resistor and by using the voltage generated across the resistor as the source of gate bias. The two basic circuits of Fig. 6.2 can be used.

Stabilisation of Depletion f.e.t.s

For depletion f.e.t.s bias and a measure of d.c. stabilisation can be obtained by including a resistor R_s in the source circuit and by returning R_g to the source supply terminal as shown in Fig. 6.15. Any changes in I_d cause corresponding changes in the voltage across R_s which are applied directly between gate and source so minimising the original change in drain current.

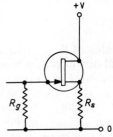

Fig. 6.15. Simple d.c. stabilising circuit which can be used with a depletion f.e.t.

The required value of R_s is determined in the following manner. Firstly the mean drain current which will give the desired performance is chosen. Next the gate bias voltage V_g which gives the chosen value of I_d is determined from the $I_d - V_g$ characteristics of the transistor. The required value of R_s is given by V_g/I_d. For example if I_d is 5 mA and V_g is 2 V, R_s is given by $2/(5 \times 10^{-3})$, i.e. 400 Ω.

Calculation of stability factor

For a field-effect transistor without stabilisation the relationship between changes of drain current i_d and changes in gate voltage v_g is simply

$$i_d = g_m v_g \tag{6.8}$$

i.e. the drain current is directly proportional to g_m for a given gate voltage and thus varies with temperature or with change of transistor in the same way and to the same extent as g_m.

For an f.e.t. with a source resistance R_s the relationship between i_d and input voltage v_{in} can be deduced in the following way:

$$v_{in} = v_{gs} + v_{fb}$$

Substituting $i_d R_s$ for v_{fb} and i_d/g_m for v_{gs} we have

$$v_{in} = \frac{i_d}{g_m} + i_d R_s$$

from which

$$i_d = \frac{g_m v_{in}}{1 + g_m R_s} \tag{6.9}$$

Comparison of Eqns 6.8 and 6.9 shows that the effect of the source resistance is to reduce variations in drain current for a given input signal to $1/(1+g_mR_s)$ of their unstabilised value. The stability factor for this circuit is thus given by

$$K = \frac{1}{1+g_mR_s} \qquad (6.10)$$

The value of R_s is, of course, fixed by bias considerations and thus the stability factor for the simple circuit of Fig. 6.15 is automatically determined. For example if R_s is 400 Ω as in the last numerical example and if g_m is 2 mA/V, the stability factor is given by $1/(1 + 2 \times 10^{-3} \times 400) = 1/1\cdot8 = 0\cdot55$. This is a very poor performance. However, greater stability can be obtained, without effect on bias, by using two resistors R_{s1} and R_{s2} in series in the source circuit and by returning R_g to their junction as shown in Fig. 6.16. Bias is

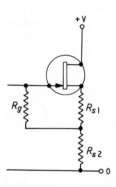

Fig. 6.16. Modification of the previous circuit to give better d.c. stabilisation than is obtainable with the normal value of automatic bias resistor

determined by the value of R_{s1} but stability by $(R_{s1}+R_{s2})$. High stability can thus be obtained by using large values for R_{s2} but such values reduce the voltage available across the transistor and its load resistor.

Minimising signal-frequency feedback

Resistors included in the source circuit give signal-frequency negative feedback (and hence reduced gain) in addition to the d.c. feedback which is responsible for the stabilising effect. Signal-frequency feedback can be minimised by decoupling the source resistors by a low-reactance capacitor as shown in Fig. 6.17: the similarity between

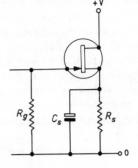

Fig. 6.17. Decoupling to minimise signal-frequency
feedback in the circuit of Fig. 6.15

this circuit and that for automatic cathode bias for a thermionic
valve is very striking.

Stabilisation of Enhancement f.e.t.s

The stabilising circuit of Fig. 6.2(b) is suitable for enhancement
f.e.t.s because it gives a gate bias voltage between that of the source
and the drain. Fig. 6.18 shows the circuit applied to an n-channel

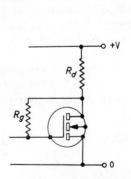

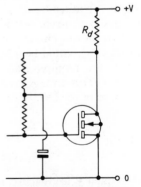

Fig. 6.18. Simple d.c. stabilising circuit for an enhancement f.e.t.

Fig. 6.19. Method of minimising signal-frequency feedback in the circuit of the
previous diagram

i.g.f.e.t. The stability factor is given by Eqn 6.10 by substituting
R_d for R_s. If the output of the transistor is taken from the drain
circuit R_g must be decoupled: this can be done as suggested in
Fig. 6.19.

Potential Divider and Source Resistor Circuit

This circuit (illustrated in Fig. 6.20) can be used for enhancement and depletion f.e.t.s and the stability factor obtained is given by expression 6.10. It is possible to use higher values of R_s than in the simple circuit of Fig. 6.15 and higher values of K are obtainable. This is illustrated in the following numerical example.

Suppose the chosen mean drain current is 1·5 mA and that the gate bias necessary to give this current is −2 V. If the supply voltage

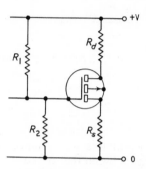

Fig. 6.20. *Potential divider and source resistor method of d.c. stabilising an f.e.t.*

is 15, a suitable value for the voltage at the junction of R_1 and R_2 is 3 V. This requires R_1 to be four times R_2. The sum of R_1 and R_2 determines the current taken by the potential divider and this can be kept low in this circuit without sacrificing stability (cf. page 87). For example R_1 could be 4 MΩ and R_2 10 MΩ.

To give the required gate-source voltage the source potential must be 5 V. This gives R_s as $5/(1·5 \times 10^{-3})$, i.e. 3·3 kΩ. If the mutual conductance of the f.e.t. is 2 mA/V the stability factor is given by

$$K = \frac{1}{1 + 2 \times 10^{-3} \times 3,300}$$

$$= 0·13$$

Signal-frequency feedback can be minimised by decoupling R_s as in Fig. 6.17.

CHAPTER 7

Small-signal A.F. Amplifiers

DEFINITION OF SMALL-SIGNAL AMPLIFIER

Some amplifiers are required to deliver an output voltage or output current which is small compared with the maximum that the amplifier could deliver. When a voltage output is wanted, the magnitude of the current output is usually of little consequence provided the transistor(s) can supply it without distortion. Similarly if the amplifier is designed to supply a current output, the magnitude of the voltage output is of secondary importance provided it is not sufficient to cause overloading and distortion. Such amplifiers are termed small-signal amplifiers: typical examples are microphone head amplifiers and the r.f. and early i.f. stages in a receiver.

DISTINCTION BETWEEN VOLTAGE AND CURRENT AMPLIFIERS

If a stage is to be regarded as a voltage amplifier, the voltage of the signal source must not be affected by the connection of the amplifier across it; thus the input resistance of the amplifier must be high compared with the resistance of the signal source. Moreover the output voltage of the amplifier must be substantially unaffected by the connection of the load: thus the output resistance must be low compared with the load resistance. Provided these resistance requirements are satisfied the voltage gains of the individual stages of an amplifier can be multiplied to give the overall voltage gain of the amplifier or (and this is another way of expressing the same fact) the voltage gains of the individual stages, when expressed in decibels, can simply be added to give the overall gain of the amplifier.

When a current amplifier is connected to a signal source the current from the source must not be affected by the connection of the amplifier: thus the input resistance of the amplifier must be small compared with the resistance of the signal source. The output current from the amplifier should not be affected by the connection of the load resistance: thus the output resistance should be high compared with the load resistance. Provided these resistance requirements are met the current gains of the individual stages of an amplifier can be multiplied to give the overall current gain of the amplifier or (and this is another way of expressing the same fact) the current gains of the individual stages, when expressed in decibels, can simply be added to give the overall gain of the amplifier.

These resistance requirements for voltage and current amplifiers are summarised in Table 7.1. From this it can be deduced that a given amplifier can behave as a voltage or current amplifier. For example if the input and output resistances of the amplifier are both 1 kΩ, then if $R_s = 100\ \Omega$ and $R_l = 10$ kΩ, the amplifier is best regarded as a voltage amplifier whereas if $R_s = 10$ kΩ and $R_l = 100\ \Omega$ it is best regarded as a current amplifier.

Table 7.1

Source and load resistance considerations
in voltage and current amplifiers

	source resistance R_s	load resistance R_l
voltage amplifier	$\ll r_{in}$	$\gg r_{out}$
current amplifier	$\gg r_{in}$	$\ll r_{out}$

In applying these theoretical considerations to practical transistor circuits, a number of precautions must be taken. For example if a bipolar transistor is fed from a low-resistance signal source (to give a voltage amplifier) severe distortion can occur for the reason given on page 44. To minimise this the source resistance must be made large compared with the input resistance: this is a reminder that the bipolar transistor is inherently a current amplifier. The resistance of an f.e.t. is so high that it is impractical to feed it from an even higher source resistance (to give a current amplifier). The input of the f.e.t. could, of course, be shunted by a low resistance but this wastes the most important property of the f.e.t. namely its high input resistance: this is a reminder that the f.e.t. is essentially a voltage amplifier.

TRANSISTOR PARAMETERS IN SMALL-SIGNAL AMPLIFIERS

It is characteristic of transistors used as small-signal amplifiers that the signal voltages at the transistor terminals are small compared with the steady potentials at these points. Similarly the signal currents have a magnitude small compared with that of the steady currents on which they are superimposed. When operation is thus confined to small excursions about a mean value, a transistor may be regarded as having substantially constant input resistance, output resistance and current gain.

SINGLE-TRANSISTOR STAGES

Bipolar Transistor

The circuit diagram for a single-stage bipolar-transistor amplifier is given in Fig. 7.1. It uses the potential divider method of d.c. stabilisation and a method of calculating values for R_1, R_2 and R_e

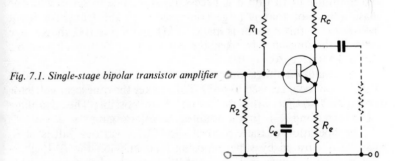

Fig. 7.1. Single-stage bipolar transistor amplifier

to give a required mean value of collector current is given on page 87. It now remains to calculate the value of R_c.

The value of R_c depends primarily on whether the amplifier is required to deliver a current or voltage output. If a current output is required, as much as possible of the output current must enter the external load resistor (shown in dashed lines). R_c must hence be large compared with this external load and in practice it is given the highest value possible consistent with maintaining an adequate steady voltage on the collector. For example if the mean collector current is 1 mA, if the base potential is 2 V and if the supply voltage is 9 V, then a suitable value for the collector potential is 3 V (giving a collector-base voltage of 1 V). There are hence 6 V across R_c and its value is $6/(1 \times 10^{-3})$, i.e. 6 kΩ. The value of R_c is not, however, critical provided

(a) it is large compared with the external resistor

(b) it does not greatly exceed 6 kΩ.

Provided the parallel resistance of R_1 and R_2 is large compared with the input resistance of the transistor and that R_c is large compared with the external load, the current gain of the amplifier is almost equal to β.

If, however, the transistor is required to give a voltage output then the external resistance will be large compared with R_c: thus R_c is effectively the load into which the transistor is operating. The precise value of R_c is now important and the value of the external resistance is non-critical provided it is large compared with R_c. The value to be given to R_c depends on the magnitude of the output voltage required. If the largest possible voltage swing is required R_c should be given the value which makes the quiescent collector voltage midway between the supply and emitter potentials. Using the values quoted for the current amplifier, the quiescent voltage should be made 6 V which permits upward and downward swings of 3 V amplitude. This fixes the value of R_c at $3/(1 \times 10^{-3}) = 3$ kΩ. To minimise distortion it is necessary to include in series with the base a resistor that is large compared with the transistor input resistance. If this resistor is made 30 kΩ and if β is 100, the voltage gain of the amplifier is, from the equation at the top of page 62, given by $100 \times 3/30$, i.e. 10.

If a smaller voltage output, say 1 V amplitude, is acceptable then R_c can be increased to 6 kΩ. This makes the quiescent collector voltage 3 V so that swings to 2 V and 4 V are possible. The advantage of so increasing R_c is that it doubles the voltage gain.

The operation of the circuit of Fig. 7.1 for various values of R_l can be illustrated by superimposing load lines on the I_c–V_c characteristics as shown in Fig. 7.2. ABC is the load line for a 3-kΩ load. It is drawn through that point on the V_c axis which corresponds to the supply voltage (9 V) to that point on the I_c axis which corresponds to the current (3 mA) which the supply voltage would drive through the load resistance. The intersections of the load line with the I_b characteristics indicate the current through the series combination of transistor and load, and the voltages across each component. For example point B shows that for a base current of 15 µA, 6 V is developed across the transistor and 3 V across the load resistor. These are, of course, the quiescent conditions required for maximum output voltage. If the input current rises to 32 µA the collector voltage falls to 3 V (point D) and if the input current falls to zero the collector voltage rises to 9 V. During amplification,

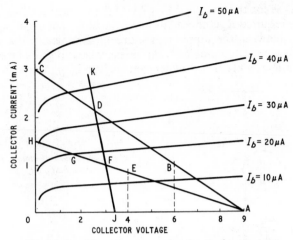

Fig. 7.2. *Load lines superimposed on $I_c - V_c$ characteristics*

therefore, the operating point swings between the limits of A and D.

If the load resistance is increased to 6 kΩ the load line becomes AEFGH and the quiescent point is at F corresponding to a collector current of 1 mA and a collector voltage of 3 V. The base current is about 17 μA and when this swings down to 14 μA and up to 19 μA the collector voltage swings up to 4 V and down to 2 V.

The operation of the current amplifier can also be represented on this diagram. The operating point is still at F because the direct-coupled collector resistor is still 6 kΩ for this amplifier. However the external load resistance (assumed capacitance-coupled) is very low and the load line JFK is therefore nearly vertical and passes through F as shown. The intercepts made on this load line by the I_b characteristics show that if the base current changes from 10 μA to 30 μA, the collector current changes from 0·6 mA to 1·9 mA, a current gain of 65 almost equal to β.

Field-effect Transistor

The circuit diagram of a single-stage amplifier using a depletion j.u.g.f.e.t. is given in Fig. 7.3. Because of the extremely-high input resistance such a stage is unlikely to be used for current amplification and we shall assume that voltage amplification is required. The method of choosing the value of R_s is described on page 95 and it remains to calculate the value of R_d. The same technique can be adopted as for the circuit of Fig. 7.1: if the largest possible voltage swing is required R_d is chosen to make the quiescent drain potential midway between the supply and source potentials but if a smaller voltage swing is acceptable R_d can be increased to give higher gain. Suppose R_d is 3 kΩ. The voltage gain is given by $g_m R_d$ and a typical

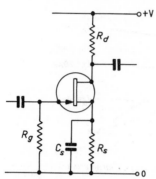

Fig. 7.3 Single-stage amplifier using a j.u.f.g.e.t.

value for g_m is 2 mA/V giving the voltage gain as $3 \times 10^3 \times 2 \times 10^{-3}$, i.e. 6. This is not a high gain and in general it is true that bipolar transistors, although inherently current amplifiers, can give higher voltage gains than f.e.t.s.

TWO-STAGE AMPLIFIERS

If a single-stage amplifier gives insufficient gain two stages can be connected in cascade and Fig. 7.4 gives the circuit diagram of a two-stage amplifier using a transformer to couple the stages, each

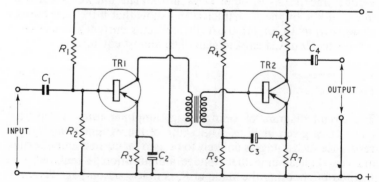

Fig. 7.4. A two-stage small-signal amplifier with transformer coupling

of which is independently stabilised by the potential divider method. Maximum gain would be obtained from the transformer if the turns ratio were chosen to match the output resistance of TR1 to the input resistance of TR2. This would, however, be a most unsatisfactory design for the following reasons:

(a) TR1 would effectively have a load resistance equal to its output resistance and this large load resistance would seriously limit its output signal amplitude (see page 102).

(b) TR2 would have an effective signal source resistance equal to its input resistance. A larger value of source resistance is necessary to minimise distortion in TR2.

(c) The transformer would need an inconveniently-large primary inductance to maintain a good bass response if TR1 has an output resistance of an appreciable fraction of a megohm as for some silicon transistors.

To avoid all these disadvantages the transformer turns ratio is kept small, e.g. 3 to 1. If the input resistance of TR2 is 1 kΩ there will be a loss of 3 dB at the frequency for which the reactance of the secondary winding equals this. If the frequency is made 50 Hz we have

$$2\pi fL = 1,000$$

$$\therefore L = \frac{1,000}{6 \cdot 284 \times 50} \text{H} = 3 \cdot 2 \text{ H approximately.}$$

If the turns ratio is 3 to 1 the primary inductance is $3^2 \times 3 \cdot 2 = 29$ H.

USE OF NEGATIVE FEEDBACK

Circuits such as that of Fig. 7.4 suffer from the disadvantage that their properties (gain, input resistance, output resistance, distortion, signal-to-noise ratio, etc.) depend on the characteristics of the transistors. It is thus difficult to make a number of amplifiers with substantially the same performance without very careful choice of transistors and components. Moreover such circuits are wasteful of components: there are simpler circuits giving better d.c. stability.

The difficulties caused by spreads in transistor parameters can be minimised by negative feedback. This technique enables the gain and frequency response to be determined by the constants of a passive network: they are then independent of transistor parameters and hence of changes in temperature. Negative feedback reduces distortion and can be arranged to increase or decrease input and output resistance as desired. The cost of these benefits is reduced gain but the required gain can be restored by the use of additional stages of amplification. The two basic circuits for negative feedback were introduced at the beginning of Chapter 6 and are reproduced in Fig. 7.5. It was shown that these are, in fact, two versions of the

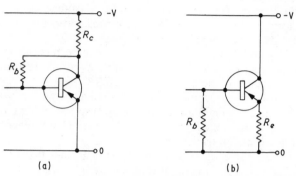

Fig. 7.5. The two basic circuits for applying negative feedback to a common-emitter amplifier

same circuit and, for a given value of R_b give identical performances in respect of d.c. stabilisation provided R_c equals R_e. D.C. feedback is, however, independent of external signal-frequency circuits. When the circuits of Fig. 7.5 are used to provide feedback at signal frequencies, account must be taken of the way in which the input

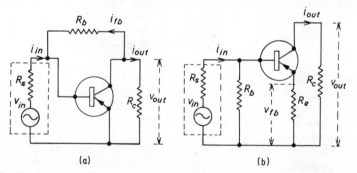

Fig. 7.6. R_b in (a) gives parallel-derived parallel-injected feedback whereas R_e in (b) gives series-derived series-injected feedback

signal is fed into the transistor and in which the output signal is taken from it. In fact, as shown in Fig. 7.6 R_b and R_e provide quite different types of signal-frequency feedback.

In Fig. 7.6(a) R_b connects the load resistor R_c and the source resistor R_s *in parallel*: in effect it introduces into the input circuit a current proportional to the output voltage. Such feedback effectively decreases the input resistance and the output resistance of the transistor. In Fig. 7.6(b) the output circuit of the transistor consists of R_c and R_e *in series* and the input circuit consists of v_{in} and R_e *in series*. R_e thus introduces into the input circuit a voltage proportional to the output current. Such feedback effectively increases the input resistance and the output resistance of the amplifier. To facilitate reference to the effects of parallel– and series–derived and –injected feedback, these are summarised in Table 7.2.

Table 7.2

Effects of negative feedback on input and output resistance

Type of feedback connection	*Effect on input resistance*	*Effect on output resistance*
series-derived		increased
parallel-derived		decreased
series-injected	increased	
parallel-injected	decreased	

Approximate expressions for the essential properties of these two fundamental feedback circuits can be deduced as follows.

Parallel-derived Circuit (Fig. 7.6(a))

The input current i_{in} is given by

$$i_{in} = i_{fb} + i_b$$

where i_{fb} is given by v_{out}/R_b. If i_{fb} is large, i_b can be neglected in comparison and we have

$$i_{in} = i_{fb} = \frac{v_{out}}{R_b}$$

from which

$$\frac{v_{out}}{i_{in}} = R_b \qquad (7.1)$$

Now $v_{out} = i_{out}R_c$. Substituting for v_{out}

$$\frac{i_{out}}{i_{in}} = \frac{R_b}{R_c} \qquad (7.2)$$

If R_s is large compared with the input resistance (as it should be to minimise distortion)

$$i_{in} = \frac{v_{in}}{R_s}$$

Substituting for i_{in} in Eqn 7.1

$$\frac{v_{out}}{v_{in}} = \frac{R_b}{R_s} \tag{7.3}$$

Eqns 7.2 and 7.3 can be derived from inspection of the circuit diagram. In Fig. 7.6(a) the output current from the transistor divides at the junction of R_c and R_b so that a fraction $R_c/(R_c+R_b)$ enters R_b and is returned to the base as feedback. Normally R_b is large compared with R_c and the current division ratio is approximately R_c/R_b. These two resistors determine the current gain of the amplifier which is given by the reciprocal of the current division ratio, i.e. by R_b/R_c.

In Fig. 7.6(a) R_b and R_s form a potential divider across the load resistor R_c and it returns to the base a fraction $R_s/(R_s+R_b)$ of the output voltage. If R_b is large compared with R_s the fraction is approximately R_s/R_b. This potential divider determines the voltage gain of the circuit which is given by the reciprocal of the division ratio, i.e. by R_b/R_s.

From Eqn 7.1 we can obtain a simple expression for the output resistance of the circuit of Fig. 7.6(a). We know that

$$i_{out} = \beta i_{in}$$

and by eliminating i_{in} between this equation and Eqn 7.1 we have

$$r_{out} = \frac{v_{out}}{i_{out}} = \frac{R_b}{\beta}$$

Thus if $\beta = 100$ and $R_b = 100$ kΩ the output resistance is 1 kΩ.

Series-derived Circuit (Fig. 7.6(b))

The signal-frequency feedback voltage v_{fb} returned to the input circuit is equal to $i_{out}R_e$ and the input voltage is given by

$$v_{in} = v_{fb} + v_{be}$$

If v_{fb} is large, v_{be} can be neglected in comparison and we have

$$v_{in} = v_{fb} = i_{out}R_e$$

from which

$$\frac{i_{out}}{v_{in}} = \frac{1}{R_e} \tag{7.4}$$

Now $v_{out} = i_{out}R_c$. Substituting for i_{out}

$$\frac{v_{out}}{v_{in}} = \frac{R_c}{R_e} \tag{7.5}$$

If R_s determines the base current $i_{in} = v_{in}/R_s$. Substituting for v_{in} in Eqn 7.4 we have

$$\frac{i_{out}}{i_{in}} = \frac{R_s}{R_e} \tag{7.6}$$

Eqns 7.5 and 7.6 can be confirmed from inspection of the circuit diagram, R_c and R_e constituting the potential divider, R_s and R_e forming the current divider.

From Eqn 7.4 we can obtain a simple expression for the input resistance of the circuit of Fig. 7.6(b). We know that

$$i_{out} = \beta i_{in}$$

and by eliminating i_{out} between this equation and Eqn 7.4 we have

$$r_{in} = \frac{v_{in}}{i_{in}} = \beta R_e \tag{7.7}$$

Thus if $\beta = 100$ and $R_e = 1$ kΩ the input resistance is 100 kΩ. In practice, of course, this may be effectively reduced by other resistors such as those of a potential divider connected to the base.

TWO-STAGE AMPLIFIERS

There are two ways in which a circuit of the type illustrated in Fig. 7.5(a) can be combined with one of the type illustrated in Fig. 7.5(b) to form a two-stage amplifier. One way is to have a first stage of type (a) with a second stage of type (b). The first stage has a low output resistance and the second has a high input resistance. These are the conditions required to transfer the output voltage of the first stage to the input of the second with little loss. The basic form of the amplifier is shown in Fig. 7.7 in which for simplicity direct coupling is employed between the stages.

If we represent the signal voltage developed across R_{c1} as v,

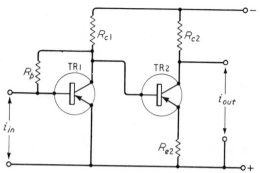

Fig. 7.7. Skeleton form of one type of two-stage amplifier

this is the output voltage of TR1 and the input voltage of TR2. For TR1 we have, from Eqn 7.1

$$\frac{v}{i_{in}} = R_b$$

For TR2 we have, from Eqn 7.4

$$\frac{i_{out}}{v} = \frac{1}{R_{e2}}$$

Eliminating v

$$\frac{i_{out}}{i_{in}} = \frac{R_b}{R_{e2}}$$

Because of the undecoupled emitter resistor R_{e2} TR2 behaves as an emitter follower and R_b can be connected to the emitter of TR2 instead of to the collector of TR1 with little effect on the performance of the circuit which now has the form shown in Fig. 7.8.

To make a practical version of this circuit we now need to add measures to ensure d.c. stability of both transistors. One of the most successful methods of stabilisation is that illustrated in Fig. 6.12 and it is easy to add to Fig. 7.8 the few components necessary to give this type of stabilisation. These include a decoupled emitter resistor R_{e1} for TR1 and a decoupled potential divider R_1R_2 in the emitter circuit of TR2. With these additions the circuit has the final form given in Fig. 7.9. The amplifier so constructed has a low input resistance and a high output resistance and is likely therefore to be used for current amplification.

The current gain of the amplifier is given by R_b/R_{e2} and the resistance values should be chosen to give the required gain. R_b

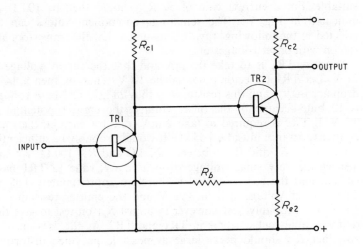

Fig. 7.8. *Modification of the circuit diagram of Fig. 7.7*

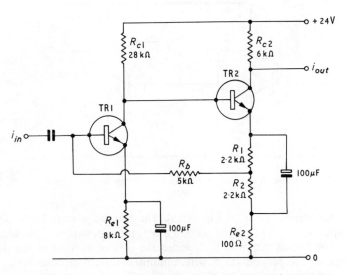

Fig. 7.9. *Practical form of the amplifier of Fig. 7.8 designed for a current gain of 50*

should be small to give good stability and a value of 5 kΩ is suitable. For a current gain of 50 R_{e2} should then be 100 Ω as indicated in Fig. 7.9. The remaining component values can be selected in the following way. It is assumed that the transistors are silicon with offset voltages of 0·7 V.

Suppose TR1 is to take 0·5 mA and that the supply voltage is 24 V. Let TR1 collector potential be 10 V. There is then a 14-V drop across R_{c1} and this resistance is thus 28 kΩ. TR2 base voltage is 10 but, because of the offset voltage, the emitter potential is 9·3 V. If TR2 is required to take 2 mA collector current, the total emitter resistance must be 4·65 kΩ. It would probably be sufficiently accurate for R_1 and R_2 to be each 2·2 kΩ and R_{e2} 100 Ω. We can neglect the very small voltage drop across R_b due to TR1 base current and the base potential is therefore approximately 4·65 V. TR1 emitter voltage is then 3·95 V and the emitter resistance is 8 kΩ. It is probably best however to adjust R_{e1} on test to give the required quiescent currents in TR1 and TR2. As this is a current amplifier R_{c2} should be as large as possible provided distortion does not occur as a result of TR2 collector potential approaching the base potential too closely. A value of 6 kΩ should be suitable.

An alternative method of obtaining a two-stage amplifier is to combine a first stage of the type illustrated in Fig. 7.5(b) with a

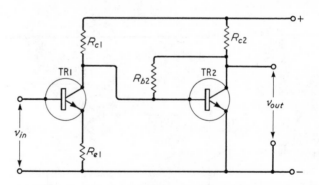

Fig. 7.10. Skeleton form of another type of two-stage amplifier

second stage similar to Fig. 7.5(a). The first stage now has a high output resistance and the second a low input resistance. These are the conditions required to transfer the output current of the first stage to the second stage with minimum loss. The basis form of the amplifier so produced is given in Fig. 7.10 in which direct coupling is employed between the stages.

If we represent the output current of TR1 and the input current of TR2 by i, we have, applying Eqn 7.4 to TR1

$$\frac{i}{v_{in}} = \frac{1}{R_{e1}}$$

For TR2 we have from Eqn 7.1

$$\frac{v_{out}}{i} = R_{b2}$$

Eliminating i between these equations

$$\frac{v_{out}}{v_{in}} = \frac{R_{b2}}{R_{e1}}$$

As shown in Fig. 7.11 it is more usual to connect R_{b2} to TR1 emitter instead of to TR2 base. This makes little difference to the performance of the circuit because the current injected by R_{b2} for the most part enters TR1 (the resistance R_{e1} being large compared with the internal emitter resistance r_e of TR1). At the collector

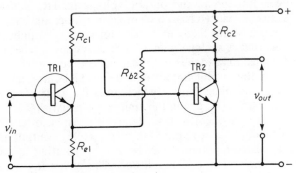

Fig. 7.11. *Modification of the circuit diagram of Fig. 7.10*

of TR1 this current mostly enters TR2 base because R_{c1} is, with proper design, large compared with TR2 input resistance. Thus the feedback current, though injected into TR1 emitter, mostly enters TR2 base and, of course, the feedback current suffers no phase inversion or current gain in TR1. To signals injected at the emitter, TR1 behaves as a common-base amplifier, the base being effectively earthed by the low resistance of the signal source.

Finally it is necessary to add means for d.c. stabilisation and once again we can employ the potential divider circuit of Fig. 6.12 as indicated in Fig. 7.12. The gain of the amplifier is given by R_{b2}/R_{e1}:

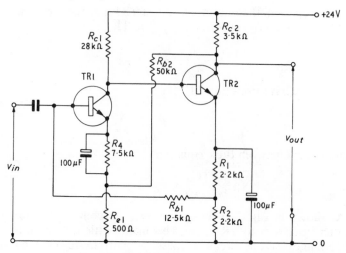

Fig. 7.12. Practical form of the amplifier of Fig. 7.11 designed for a voltage gain of 100

these resistors constitute a potential divider across the output circuit returning a fraction of the output voltage to TR1 emitter. The amplifier so constructed has a high input resistance and a low output resistance and is thus likely to be used for voltage amplification.

Suppose that a voltage gain of 100 is required. The amplifier can be designed to have the same mean collector currents and supply voltage as for the current amplifier. As before TR1 quiescent collector potential can be taken as 10 V which gives R_{c1} as 28 kΩ. TR2 base potential is also 10 V, giving the emitter potential as 9·3 V (silicon transistors are assumed) so that R_1 and R_2 can each be 2·2 kΩ as before. The amplifier is required to give a voltage output and TR2 quiescent collector potential can be 17 V permitting upward and downward swings of 7 V amplitude. This gives R_{c2} as 3·5 kΩ. R_{b2} should not unduly shunt R_{c2} (or the supply!) and a suitable value is 50 kΩ. For a voltage gain of 100, R_{e1} must hence be 500 Ω. The total resistance in TR1 emitter circuit should be 8 kΩ to give the required quiescent emitter potential and R_4 should hence be approximately 7·5 kΩ. R_4 should preferably be made adjustable (with a range of say 5 kΩ to 10 kΩ) to enable the desired working voltages and currents to be set up. R_{b1} should be small to give good d.c. stability but, as it shunts the input to TR1, it should be large to give a high input resistance to the amplifier. The input resistance of TR1 is given by βR_{e1} (Eqn 7.7), i.e. 50 kΩ if β is 100. To keep the amplifier input resistance above 10 kΩ, R_{b1} should not be less than 12·5 kΩ.

COMPLEMENTARY AMPLIFIERS

There are advantages in using complementary transistors in a two-stage amplifier. For example this simplifies the provision of negative feedback as shown in the circuit diagram of Fig. 7.13. The npn common-emitter stage TR1 is direct-coupled to the pnp common-emitter stage TR2. The potential divider $R_{c2}R_{e1}$ is the collector

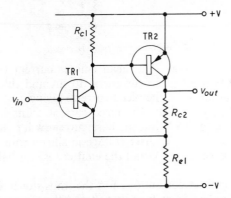

Fig. 7.13. Simple two-stage amplifier using complementary transistors

load of TR2 and returns the fraction $R_{e1}/(R_{e1} + R_{c2})$ of the output voltage to TR1 as negative feedback. This parallel-derived series-injected feedback gives the amplifier a high input resistance and a low output resistance so that the amplifier is best suited to voltage amplification. The voltage gain is given by $(R_{c2} + R_{e1})/R_{e1}$ and if R_{c2} is large compared with R_{e1}, as is likely, the gain is given approximately by R_{c2}/R_{e1}.

EMITTER FOLLOWER

An emitter follower is often used as the first stage of a voltage amplifier to provide the required high input resistance. Such a stage is also often employed as the final stage of a voltage amplifier to give a low output resistance.

A typical emitter follower circuit is given in Fig. 7.14 and the resistor values can be estimated in the following way. The aim is to provide a high input resistance and so a good starting point is to make R_1 and R_2 equal and to give them a reasonably high value such as 100 kΩ. This will ensure an input resistance of 50 kΩ. Suppose a 15-V supply is available and that the transistor is a

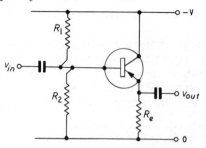

Fig. 7.14. Simple emitter follower circuit

silicon type with $\beta = 100$. If a mean emitter current of 1 mA is assumed then the mean base current is 10 μA and this will flow through R_1 (in addition to the bleed current) giving a voltage drop of $10 \times 10^{-6} \times 10^5$, i.e. 1 volt. The bleed current on its own would give a base voltage of 7·5 V but the base current will reduce this to 6·5 V. Because of the 0·7-V offset voltage in silicon transistors, the emitter voltage is hence 5·8 V and the emitter resistor value is thus 5·8 kΩ.

With a mean emitter current of 1 mA the transistor has a mutual conductance of approximately 40 mA/V and the output resistance of the circuit is given by $1/g_m$, i.e. 25 Ω.

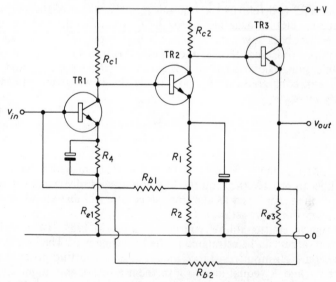

Fig. 7.15. Illustrating the addition of a direct-coupled emitter follower to the output of the circuit of Fig. 7.12

The stability of the circuit can be calculated from Eqn 6.6 and is approximately 0·08. Though satisfactory this is not as good as can be achieved by the potential divider circuit but is limited in this circuit by the need to keep R_1 and R_2 high to achieve the required high input resistance.

A circuit similar to that of Fig. 7.14 could be used in an emitter follower at the output of an amplifier but it is simpler to use a direct-coupled circuit such as that shown in Fig. 7.15. This is based on the circuit of Fig. 7.12 with an emitter follower stage added. The signal-frequency feedback resistor is taken from TR3 emitter at which the signal voltage is equal to that at TR2 collector.

DARLINGTON CIRCUIT

A simple and convenient method of connecting two transistors in cascade is the Darlington or super-alpha circuit shown in Fig. 7.16. TR1 is an emitter follower and TR2 a common-emitter amplifier.

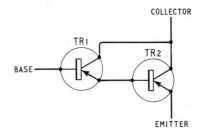

Fig. 7.16. Basic connections for the Darlington or super-alpha circuit

The collector current of TR1 is in phase with that of TR2 and thus the two collectors can be bonded as shown. The current gain of the emitter follower and the common-emitter amplifier are both approximately equal to β and thus the current gain of the Darlington circuit is given by $\beta_1 \beta_2$. The emitter current of TR1 is the base current for TR2: it follows that the collector current of TR2 is β_2 times that of TR1. If TR1 is to have a useful value of β_1 the collector current for TR1 should not be too small. TR2 should preferably therefore have a collector current of at least some tens of mA. This circuit arrangement is well suited for applications where TR2 has to supply appreciable output power.

The combination can be regarded as a single transistor with a very high value of β and consequently a high input resistance.

Where more gain is required than is available without undue sacrifice of feedback from the two-stage amplifiers of Figs. 7.9 and 7.12 a convenient method of increasing gain is to replace one of

the transistors by a Darlington pair. As an example Fig. 7.17 gives
the circuit diagram of a current amplifier using a Darlington pair
as a second stage. Such an amplifier should easily be capable of a
gain of 1000 whilst still retaining considerable feedback. The

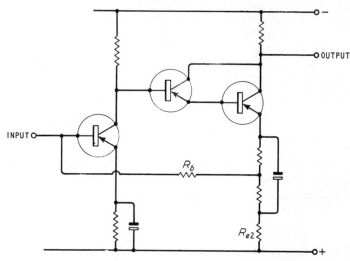

Fig. 7.17. A current amplifier employing a Darlington circuit

values of R_b and R_{e2} should be chosen to give the required value
of gain.

LOW-NOISE TRANSISTOR AMPLIFIERS

For certain a.f. amplifiers the input signal is very small and, to
obtain a good signal-to-noise ratio at the amplifier output, care
must be taken to minimise noise generated in the amplifier itself.
An example of such an amplifier is one intended to follow a high-
quality microphone. The first stage of an amplifier is, of course,
the most likely source of noise and here it is advisable to use a
transistor specially selected for use as a low-noise amplifier. It is
possible by careful selection of a bipolar transistor and by optimum
choice of operating conditions to achieve a noise factor as low as
2 dB. The noise factor is a direct measure of the added noise: thus if
the signal-to-noise ratio is 50 dB at the input to an amplifier and
47 dB at the output, the noise factor of the amplifier is 3 dB.

F.e.t.s are however preferable for use in low-level stages. Firstly
because they use only one type of charge carrier they tend to be

quieter than bipolar transistors which employ both types of charge carrier. J.u.g.f.e.t.s are in general quieter than i.g.f.e.t.s. Secondly the I_d–V_g characteristics of f.e.t.s are not linear (they are of square-law shape) so that to minimise distortion the amplitude of the input signal should be kept low. In analogue equipment, therefore, f.e.t.s should be confined to the early stages where signal levels are lowest.

CIRCUITS EMPLOYING F.E.T. AND BIPOLAR TRANSISTORS

For the reasons just given an f.e.t. is often used as the input stage of a.f. equipment and is followed by a bipolar transistor. A typical circuit is illustrated in Fig. 7.18. TR1 is a p-channel enhancement f.e.t. used as a common-source amplifier feeding into an npn emitter

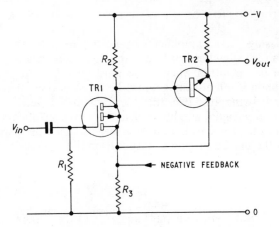

Fig. 7.18. An f.e.t. in the input stage of an a.f. amplifier

follower stage TR2. The input resistance is determined by R_1 which can be given a high value such as 1 MΩ to make the amplifier suitable for following crystal pickups or microphones. The combination of TR1, TR2 and R_2 is available as an integrated circuit.

CHAPTER 8

Large-signal A.F. Amplifiers

DEFINITION OF A LARGE-SIGNAL AMPLIFIER

The final stage of an amplifier may be required to drive a loud-speaker, a recording head or some other load requiring appreciable power for its operation. Such stages must give the required power and to obtain it full advantage must be taken of the voltage swing and current swing available from the transistor(s). Such stages are known as large-signal amplifiers. There is a danger of overloading and one of the problems in designing large-signal amplifiers is how to obtain the maximum power output from the transistors without distortion.

TRANSISTOR PARAMETERS IN LARGE-SIGNAL AMPLIFIERS

Because of the very large current swings in a large-signal amplifier it is not possible to assume, as in a small-signal amplifier that the transistor parameters are constant. The input resistance, output resistance and current gain all depend on emitter current and in a large-signal amplifier the variations of these parameters which occur during each cycle of input signal can be very great. Design of such stages must hence be carried out in terms of the mean value of the input resistance, etc., or, better, by evaluation of the extreme values of each parameter and ensuring that the desired performance can be obtained even at the extreme values.

CLASS-A AMPLIFIERS

The circuit of a class-A common-emitter output stage is given in Fig. 8.1, and a typical set of $I_c - V_c$ characteristics is given in Fig.

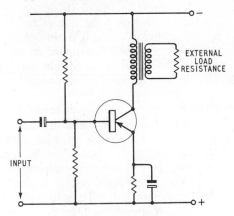

Fig. 8.1. Essential features of a single-ended class-A transistor output stage

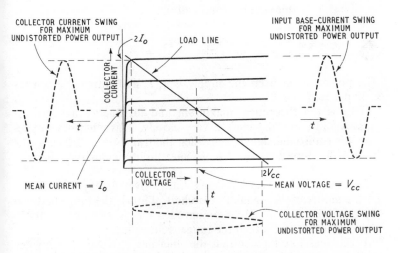

Fig. 8.2. Conditions in a class-A transistor output stage

8.2. The problem is to obtain maximum undistorted power from the transistor without exceeding the maximum safe dissipation prescribed by the manufacturers.

Efficiency

Suppose the supply voltage is V_{cc} and the steady collector current in the absence of an input signal is I_o. Then the power taken from the supply is $V_{cc}I_o$ and most of this is dissipated as heat in the transistor. This static dissipation must not exceed the maximum value quoted by the manufacturers. If a sinusoidal input signal is applied to the transistor, power is supplied to the load but the power taken from the supply remains constant because no change has taken place in the average or d.c. component of the collector current.

It follows that the power dissipated in the transistor becomes less when the input signal is applied. If, therefore, the transistor does not become too hot in the absence of an input signal, there will be no danger at all of damaging it by heat when the signal is applied.

When the transistor is delivering its maximum undistorted power, the peak value of the collector voltage is nearly equal to V_{cc} and the peak collector current is I_o. The output power is obtained by multiplying the r.m.s. collector voltage by the r.m.s. collector current: for a sinusoidal output signal these are $V_{cc}/\sqrt{2}$ and $I_o/\sqrt{2}$ and thus

$$\text{maximum power output} = \frac{V_{cc}}{\sqrt{2}} \cdot \frac{I_o}{\sqrt{2}}$$

$$= \frac{V_{cc}I_o}{2}$$

The power taken from the supply is $V_{cc}I_o$ and thus the efficiency is 50 per cent.

This is the theoretical maximum and in practice a transistor class-A stage can approach it very closely. This efficiency applies only for sinusoidal signals and when the transistor is driven to the limit of its output power. For smaller input signals the efficiency is less.

The amplitude of an a.f. signal varies over a range depending on the nature of the signal. For orchestral music the range between the maximum and minimum amplitudes may be as much as 40 dB. For pop music and speech the range is much less.

It can therefore be seen that the efficiency of a class-A amplifier with an a.f. input varies from instant to instant and the average efficiency is in practice much less than the theoretical maximum of 50 per cent.

Optimum Load

As can be seen from Fig. 8.2 the slope of the load line for maximum power output is given by V_{cc}/I_o: this is then the value of the optimum load resistance.

As a numerical example, consider a transistor rated for 100 mW maximum collector dissipation and operating as a class-A output stage from a 6-V supply.

The maximum undistorted output power is 50 mW and the mean collector current I_o is given by

$$I_o \times 6 = 100 \times 10^{-3}$$

$$\therefore I_o = \frac{100 \times 10^{-3}}{6} \text{ A}$$

$$= 17 \text{ mA}$$

$$R_l = \frac{V_{cc}}{I_o}$$

$$= \frac{6}{17 \times 10^{-3}} \Omega$$

$$= 350 \Omega$$

One practical point is that the supply voltage must exceed 6 V to give a 6-V swing of collector potential. This is because the steady voltage across the emitter resistor, that across the primary resistance and the transistor knee voltage must be subtracted from the supply voltage to give the effective collector-emitter voltage.

Output Transformer

The output transformer must match the optimum load to the load resistance. Therefore, if the load resistance is 3 Ω, the transformer must have a ratio of

$$\sqrt{(350/3)} : 1 = \sqrt{117} : 1$$

$$= 11 : 1, \text{ approximately}$$

The primary inductance determines the low-frequency response which is 3 dB down at the frequency for which the inductive reactance is equal to the optimum load. In an a.f. amplifier the 3 dB loss frequency may be 50 Hz and we have

$$2\pi f L = R$$

$$L = \frac{R}{2\pi f}$$

$$= \frac{350}{6\cdot28 \times 50} \text{ henrys}$$

$$= 1\cdot1 \text{ henrys}$$

The transformer should have a primary inductance of this value with 17 mA direct current flowing.

Push-pull Operation

Two transistors may be operated in class-A push-pull and a typical circuit is given in Fig. 8.3. The collector-to-collector optimum load is twice that for a single class-A transistor and the output transformer needs twice the primary inductance. Its design is

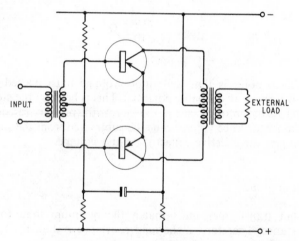

Fig. 8.3. Circuit for a push-pull class-A transistor output stage

simpler, however, because it is not polarised, the d.c. components of the collector currents flowing in opposite directions in the two halves of the primary winding. There is, however, considerable dissipation in a class-A stage and if push-pull operation is used, advantage may be taken of the cancellation of distortion by biasing back the transistors to obtain class-B operation.

CLASS-B AMPLIFIERS

In a class-B amplifier the base is biased almost to the point of collector current cut-off. In the absence of an input signal, therefore, very little collector current flows but the current increases as the amplitude of the input signal increases. This leads to economy in running costs because the current taken from the supply tends to be proportional to the signal amplitude and not independent of it as in a class-A amplifier. Alternate half-cycles of the input signal are, however, not reproduced in a single class-B amplifier and it is essential to use two transistors in push-pull. Such a pair of transistors can give a power output of 2·5 times the maximum permissible collector dissipation of the two transistors: this is 5 times the power output available from the same two transistors operating in class-A push-pull. This may be shown in the following way. Let the peak collector current of each transistor be I_p and the supply voltage V_{cc}. Then the peak collector-voltage swing V_p is equal to V_{cc}. The power output from a push-pull stage is given by the product of the r.m.s. collector current ($I_p/\sqrt{2}$) and the r.m.s. collector voltage ($V_p/\sqrt{2}$) and is thus given by $I_p V_p/2$. If the input

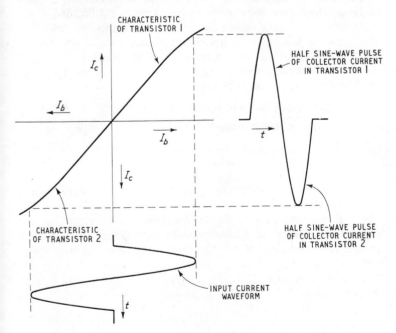

Fig. 8.4. Operation of a push-pull class-B transistor amplifier

signal is sinusoidal, the combined collector current is sinusoidal in waveform as pictured in Fig. 8.4. Such a waveform has a mean value, or d.c. component, equal to $2/\pi$ of the peak value (I_p) and the power taken from the supply is thus equal to $2V_{cc}I_p/\pi$, giving efficiency as

$$\frac{\text{power output from transistors}}{\text{power taken from supply}} = \frac{I_p V_p}{2} \cdot \frac{\pi}{2I_p V_p}$$

$$= \frac{\pi}{4}$$

$$= 78 \cdot 54 \text{ per cent}$$

This is the maximum efficiency of which the stage is capable and to obtain it the optimum load per transistor must be equal to (peak voltage)/(peak current). For class-B push-pull the collector-to-collector load is given by $4V_p/I_p$.

If the load and supply voltage are kept constant and the input signal amplitude is reduced the efficiency falls linearly due to the increasing failure to make use of the voltage swing available. This is easily shown. Suppose the input voltage is reduced to a times the value which gives maximum output. Then the collector-current swing falls to aI_p and (since the load is constant) the collector-voltage swing falls to aV_p, giving the power output per pair of transistors as $a^2 I_p V_p/2$. The mean value of the collector current is now $2aI_p/\pi$ and the power taken from the supply is $2aI_p V_p/\pi$. The efficiency for the reduced input signal is given by

$$\frac{\text{power output from transistors}}{\text{power taken from supply}} = \frac{a^2 I_p V_p}{2} \cdot \frac{\pi}{2aI_p V_p}$$

$$= a \cdot \frac{\pi}{4}$$

The efficiency is thus directly proportional to the input signal amplitude.

In transistor amplifiers we are particularly interested in the power P_t dissipated in the transistors themselves, for it is this which causes heating of the junctions and can damage them. Now

$$\begin{array}{ccc} \text{power dissipated} & = & \text{power taken} & - & \text{power output} \\ \text{in transistors} & & \text{from supply} & & \end{array}$$

$$\text{i.e. } P_t \quad = \quad \frac{2aI_p V_p}{\pi} \quad - \quad \frac{a^2 I_p V_p}{2}$$

Differentiating this

$$\frac{dP_t}{da} = \frac{2I_pV_p}{\pi} - aI_pV_p$$

Equating this to zero to find the maximum,

$$a = \frac{2}{\pi}$$

The heat in the transistors is thus a maximum when the signal amplitude is $2/\pi$ (approximately 0·63) times that giving maximum output power. Substituting this particular value of a in the general expression given above we have

$$\text{power output from amplifier} = \frac{a^2I_pV_p}{2}$$

$$= \frac{2I_pV_p}{\pi^2}$$

$$\text{power taken from supply} = \frac{2aI_pV_p}{\pi}$$

$$= \frac{4I_pV_p}{\pi^2}$$

By subtraction,

$$\text{power dissipated in transistors} = \frac{2I_pV_p}{\pi^2}$$

These results show that this particular value of a makes the power output one-half that taken from the supply: in other words it coincides with an efficiency of 50 per cent. The maximum power output from the class-B pair is $I_pV_p/2$. The maximum power dissipated in the transistors is $2I_pV_p/\pi^2$. The ratio of these two quantities is $\pi^2/4$, approximately 2·5:1, showing that it is possible to obtain an undistorted output of 2·5 times the rated maximum collector dissipation of the two transistors, i.e. 5 times the maximum collector dissipation of one of them. This applies only for a sinusoidal input.

As a numerical example, consider two transistors each with a maximum collector dissipation of 100 mW. In class-B push-pull it is possible to obtain from these an output power of 500 mW. If the supply voltage is 6 V, this is also the peak value of the collector voltage, and the peak current is given by I_p where

$$P = \frac{1}{2} V_p I_p$$

$$\therefore I_p = \frac{2P}{V_p}$$

$$= \frac{2 \times 500}{6} \text{ mA}$$

$$= 170 \text{ mA approximately}$$

The collector-to-collector load is given by

$$R_l = \frac{4V_p}{I_p}$$

$$= \frac{4 \times 6}{170 \times 10^{-3}} \Omega$$

$$= 140 \ \Omega \text{ approximately}$$

Driver Stage

The transistor feeding a class-B output stage may be regarded as a large-signal amplifier because it must supply appreciable power to the base circuits of the output stage. The design of the driver stage and the transformer coupling it to the output stage depends on the input resistance of the output stage.

It was mentioned at the beginning of this chapter that the transistor parameters cannot be taken as constant in large-signal amplifiers. This is particularly true of the input resistance of a class-B amplifier. Such an amplifier is biased almost to cut off of collector current in the absence of an input signal and its input resistance is then high. On peaks of input signal, however, the amplifier is driven to large collector currents and at these instants the input resistance is low. The variation may be from, say, 2 kΩ for small signals to less than 100 Ω for large signals.

We have seen in earlier chapters that to obtain an undistorted input current in such circumstances the transistors must be driven from a high-resistance source. If too low a source resistance is used, the input current tends to be too low at low values of input. Thus a sine-wave input signal is reproduced with a waveform similar to that shown in Fig. 8.5. This shows symmetrical distortion: such a waveform is typical of odd-order harmonic distortion, i.e. 3rd, 5th, 7th, etc., and is particularly objectionable – more so than even-order harmonic distortion. Fig. 8.5 represents the input and output waveforms for an amplifier giving *cross-over distortion*, so called because it occurs when one transistor is being cut off and the other turned on, i.e. when the state of conduction is being transferred from one transistor to the other.

Unfortunately transistors giving appreciable power output often have significant fall off in current gain at high collector currents and this can cause distortion of the output when the signal source is of high resistance. To minimise distortion due to this cause the

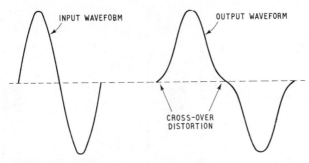

Fig. 8.5. Input and output waveforms for an amplifier giving cross-over distortion

source resistance should not be too great: on the other hand it should not be too low otherwise cross-over distortion occurs. Between these limits there is a range of source resistance which is satisfactory. Suitable values for a particular class-B amplifier are quoted below.

For transistors rated for 100 mW maximum collector dissipation, the input resistance falls to approximately 100 Ω at minimum and is probably around 1 kΩ for a standing collector current of 2 mA, a typical value for a class-B output stage for a portable receiver. If the driver stage operates in class-A with 3-mA mean collector current then for a 6-V supply the optimum load is approximately 2 kΩ. If this is accurately matched to 100 Ω we can ensure that the driver delivers maximum power to the output stage just when it is needed. The turns ratio required is given by $\sqrt{(2,000/100)} : 1 = 4\cdot5 : 1$. The effective source resistance (if the driver output resistance is 30 kΩ) is then 1·5 kΩ which is certainly not large compared with the maximum input resistance (1 kΩ) of the output stage for small inputs. Any cross-over distortion which occurs can probably be minimised by reducing the transformer turns ratio to 3:1 or even 2:1. For 2:1 the effective source resistance is 7·5 kΩ.

SYMMETRICAL CLASS-B AMPLIFIER

The circuit of the driver stage and class-B output stage of an amplifier is given in Fig. 8.6. The driver stage operates in class A and the collector current is stabilised at, say, 3 mA by the potential-divider method. A similar technique is applied to the output stage which has a common-emitter resistance, the potential divider being adjusted

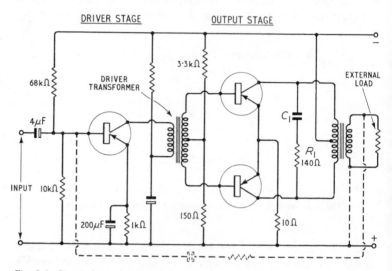

Fig. 8.6. Circuit for a driver and a symmetrical class-B output stage. Dotted lines show how negative feedback can be applied when the driver stage has a current input

to give the required standing collector current. The emitter resistance of a class-B amplifier cannot be decoupled because the mean-emitter current is not constant (as in a class-A amplifier) but varies over a wide range depending on the instantaneous amplitude of the input signal. A capacitor in parallel with the emitter resistor would acquire charge on current peaks and discharge in the intervals between peaks, thus introducing unwanted exponential signals into the amplifier. The emitter resistance must therefore be kept small to minimise the negative feedback which would reduce the gain of the output stage.

The emitter resistance is commonly as low as 5 Ω: nevertheless, it gives some measure of stabilisation of the standing collector current and it helps to equalise differences in parameters of the two output transistors. Another resistance which must be small is the lower arm of the potential divider feeding the bases of the output transistors. The input currents of the output stage flow in this resistor and a large value would severely reduce the signal input to the output stage: a value as low as 120 Ω is commonly used. It is not advisable to decouple this resistance by a low-reactance capacitor because on large input signals the capacitor becomes charged by the unidirectional input current flow for the output stage and on small input signals it discharges through the resistance, thus also superimposing unwanted exponential signals on the input to the output stage.

Class-B output stages tend to give unwanted ripples on the output signal: these are caused by ringing of the secondary winding of the driver transformer when the associated transistor is cut off. This tendency can be reduced by so constructing the transformer that the two secondary windings are closely coupled, e.g. by winding the secondaries in a bifilar manner. This discourages ringing because the undamped secondary winding is effectively damped by its close coupling to the damped winding. The network $R_1 C_1$ is included to attenuate unwanted high-frequency components such as harmonics of the input signal or oscillations due to ringing. This network together with the leakage inductance L of the output transformer and the resistance R_l of the external load form an LCR combination. Such a network can give an input resistance independent of frequency if the product $R_1 R_l$ is made equal to L/C.

Distortion introduced by the class-B output stage can be reduced by negative feedback. The feedback voltage can be taken from the secondary winding of the output transformer and applied to the driver stage. It should be applied to the base if the driver stage is required to operate with a current input, and to the emitter if the driver stage is required to operate with a voltage input. The latter alternative is preferable if the driver stage directly follows a detector, as may occur in a receiver. The degree of feedback which can be applied is limited by the phase shifts introduced by the driver and output transformers at the extremes of the passband and in practice rarely exceeds 6 dB: even this small degree halves the harmonic distortion of the amplifier.

ASYMMETRIC CLASS-B AMPLIFIER

An interesting variant of the class-B push-pull circuit is the so-called single-ended or asymmetric circuit illustrated in Fig. 8.7. The output transistors are connected in series across the collector supply battery and the load resistance is connected between their junction and a centre tap on the battery. Separate secondary windings are required for the two bases because there is now a difference in the steady voltage of the two. The optimum load equals (peak collector voltage

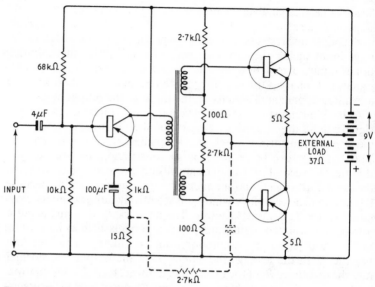

Fig. 8.7. Circuit for a driver and asymmetric class-B output stage. Dotted lines show how negative feedback can be applied when the driver stage has a voltage input

swing)/(peak collector current swing), and for output transistors operating from a 4·5-V supply and taking a peak collector-current of 120 mA may be as low as 37 Ω. A loudspeaker speech coil can readily be wound to have an impedance of this value and thus there is no need for a loudspeaker matching transformer: this is one of the great advantages of this circuit. It does, however, require a collector supply of higher voltage than a symmetrical circuit and the input transformer must have isolated secondary windings. The elimination of the output transformer is a great asset, however, and makes this circuit particularly suitable for miniature receivers. The use of a centre-tapped battery (as in Fig. 8.7) is not essential. The external load may alternatively be returned to the positive battery terminal (usually at earth potential) provided a capacitor is connected in series with the load to avoid disturbance of d.c. conditions. In an a.f. amplifier employing a 37 Ω load, the capacitance should be not less than 100 μF to avoid loss of bass.

COMPLEMENTARY CLASS-B AMPLIFIER

In the circuit of Fig. 8.7 the output transistors must be driven by antiphase signals to obtain push-pull action: thus a transformer or some other phase-splitting device is needed to provide suitable

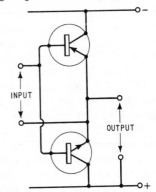

Fig. 8.8. Basic form of complementary single-ended push-pull output stage

input signals. An alternative possibility is to use two complementary transistors as an output stage as suggested in the simplified diagram of Fig. 8.8. This circuit has the advantage that no phase splitter is necessary. A positive-going signal applied to the base of the pnp transistor reduces its collector current and at the base of the npn transistor increases collector current. The pnp transistor is in fact turned off and the npn transistor is turned on: thus the potential of their common connection point moves towards that of the positive supply terminal. Similarly a negative-going input signal causes the potential of the common point to move towards the negative supply terminal. Both transistors contribute towards the output voltage and push-pull action is obtained by applying the same input signal to both bases. This basic circuit has something in common with the complementary emitter follower of Fig. 9.10 but in Fig. 8.8 the transistors are used as common-emitter amplifiers by applying the common input between base and emitter of both transistors.

A complementary class-B output stage of this basic type can be used to supply up to approximately 1 W but it is not easy to find a pair of accurately matched pnp and npn transistors to supply 10 W or more. This difficulty can be overcome by using matched output transistors of the same type and making the driver stage of complementary transistors as shown in the simplified diagram of Fig. 8.9. TR3 is an emitter follower: its output is in phase with its input and is connected directly between base and emitter of TR5. TR4 has the same input as TR3 but is a common-emitter amplifier: its output is in antiphase with the input and is connected directly between base and emitter of TR6. We have seen earlier that the current gain of the emitter follower and the common-emitter amplifier are both approximately equal to β. Thus by choosing complementary transistors with equal βs for TR3 and TR4, and by making R_{10} equal to R_{12} we can arrange for TR5 and TR6 to receive equal-amplitude but antiphase signals so that the output stage operates in push-pull.

A technique commonly employed to drive TR3 and TR4 is indicated in Fig. 8.10 which gives the circuit diagram with most component values for an amplifier capable of supplying an output of the order of 10 W. TR2 is a common-emitter pnp stage, R_7 is its collector load and R_6 a decoupling resistor. The decoupling capacitor C_3 is, however, returned to the emitters of TR3 and TR4: this ensures that the signals generated by TR2 are applied between base and emitter of TR3 and TR4 as required. R_9 determines the standing voltages on the bases of TR3 and TR4 which in turn control the collector current of the output stage: R_9 is adjusted to give a standing current in TR5 and TR6 sufficient to minimise cross-over distortion. The forward-biased diode D1 in conjunction with the emitter resistors R_{13} and R_{14} stabilise the quiescent current of the output stage as explained in Chapter 6.

TR1 is a common-emitter npn stage used for signal amplification, the collector load R_3 being connected directly between TR2 base and emitter. TR1 is also used as a d.c. amplifier to ensure stability of the quiescent voltage at the amplifier output point P. This voltage must be accurately maintained at $V_{cc}/2$ where V_{cc} is the supply voltage, otherwise the amplifier cannot deliver its maximum output without distortion. The potential at point P is applied to TR1 emitter via R_5, a.f. signals being attenuated by C_2. TR1 base is fed from the potential divider $R_1 R_2$ which provides a constant reference potential. Any variation in the quiescent voltage at point P is amplified by TR1 and the subsequent stages, all of which are direct coupled, so as to minimise the change in potential at P.

Signal-frequency negative feedback is provided by R_{15} and R_4 which return a fraction of approximately 1/9th of the output voltage to TR1 emitter. This means that the overall voltage gain of the amplifier is approximately 9. As the maximum excursion of output voltage is 22·5 V (i.e. $V_{cc}/2$) the input required is approximately 2·5 V peak. C_4 corrects the negative feedback for phase shifts in the amplifier at high frequencies and so avoids a peak in the frequency response curve or possible instability at or above the upper limit of the passband.

In general the maximum peak output voltage is $V_{cc}/2$ and thus the r.m.s. voltage is $V_{cc}/2\sqrt{2}$. If the load resistance is R_l the maximum power output is given by V^2/R_l, i.e.

$$P_{max} = \left(\frac{V_{cc}}{2\sqrt{2}}\right)^2 \cdot \frac{1}{R_l}$$

$$= \frac{V_{cc}^2}{8R_l}$$

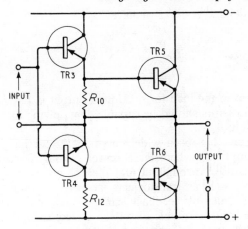

Fig. 8.9. A single-ended push-pull stage with a complementary driver stage

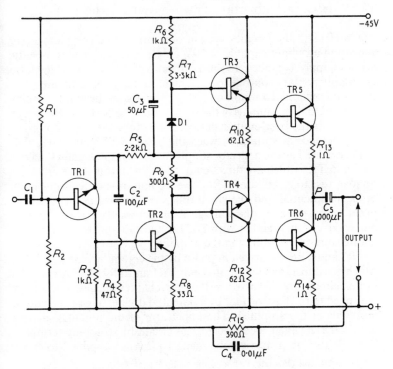

Fig. 8.10. Circuit diagram of an amplifier capable of more than 10 W output

For $V_{cc} = 45$ and $R_l = 15$ we have

$$P_{max} = \frac{45^2}{8 \times 15} \text{ W}$$

$$= 17 \text{ W}$$

The current in the load for maximum output is 1 A r.m.s. and the output transistors should be capable of supplying this without too great a fall in β.

This circuit has been widely used for high-fidelity and public address amplifiers where large power output is required. Direct coupling is used throughout and there is no need for iron-cored components such as transformers, not even a loudspeaker transformer. Considerable feedback can be used and it is possible to reduce the total harmonic distortion to less than 1%. The circuit was first described by Tobey and Dinsdale.*

INTEGRATED-CIRCUIT A.F. POWER AMPLIFIER

Fig. 8.11 gives the circuit diagram of a monolithic complementary push-pull amplifier capable of 1 W output and designed to operate with a supply balanced about earth potential. TR7 and TR8 form a complementary output pair which, unlike the amplifier just described, share a common-collector connection, the emitters being connected to the terminals of the supply. Because of the small voltage drop across the base-emitter path of transistors, there is a considerable difference in voltage between the bases of TR7 and TR8. This voltage is used as the collector supply for the complementary drivers TR5 and TR6 which are direct coupled to the output stage and have no load resistors. TR5 and TR6 have a common-emitter connection at earth potential and the small difference in voltage between the bases is provided by the voltage drop across the forward-biased diode D1 which, together with D2, D3 and the transistors TR3 and TR4 ensures d.c. stability in the amplifier.

TR5 and TR6 are driven in phase by the common-base stage TR3 which is fed from the long-tailed pair TR1 and TR2. The advantages of this circuit are discussed in the next chapter. Negative feedback is applied to the base of TR2 by a potential divider across the output terminals of the amplifier. The upper arm of the potential divider is R_5 (10 kΩ) which is connected into circuit by strapping terminals 5 and 7: the lower arm can be made 1 kΩ by strapping 8 to earth,

* Tobey, R. and Dinsdale, J. 'Transistor Audio Power Amplifiers', *Wireless World*, **67**, No. 11, 565 (1961).

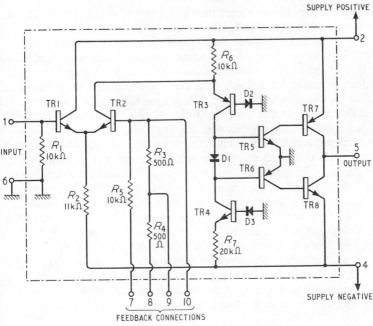

Fig. 8.11. *Circuit diagram of a monolithic a.f. power amplifier*

500 Ω by strapping 9 to earth or 250 Ω by strapping 8 and 10 and earthing 9. Thus there are three possible values of voltage gain and from inspection of the component values we can see that these are approximately 10, 20 or 40.

The amplifier has an input resistance of approximately 8·5 kΩ and an output resistance of approximately 0·5 Ω. A supply of ±6 V can be used and 1 W can be delivered to a 16-Ω load. The amplifier is contained in a can about $\frac{1}{3}$-in in diameter which should preferably be cooled by a heat sink.

HEAT SINKS

As we have seen, transistors used in output stages are often required to dissipate appreciable power in the transistors themselves. In a class-A stage this internal dissipation is a maximum at the instants when there is no input signal: the power given by the product of the standing collector current and the collector-emitter voltage is then entirely dissipated in the transistor. In a class-B stage with a sinusoidal input signal, dissipation in the transistors is a maximum when the input-signal amplitude is $2/\pi$ (approximately 0·63) of that for maximum power output.

With such dissipations there is a real danger of thermal runaway and it is advisable to provide some method of removing heat from the transistors. The usual method is to mount the transistors on a sheet of metal, known as a heat sink, which should have adequate thermal capacity (mass × specific heat) and the transistors are usually so designed that it is easily possible to bring the collector electrode into good thermal contact with the heat sink.

The primary limitation on the power which can safely be dissipated in a transistor is set by the temperature within the semiconducting material of the transistor. This is known as the collector junction temperature T_j and its maximum value is decided after life tests on the transistor. Operation of the transistor above this maximum value may seriously damage or even destroy the transistor. The circuit designer is not required to measure the junction temperature but it is helpful if he can estimate it. A method of doing this is suggested below.

The junction temperature of a transistor is always higher than that of the air immediately adjacent to the transistor case (the ambient temperature T_{amb}) the difference being proportional to the power dissipated in the transistor. Thus

$$T_j - T_{amb} = P_c \times \text{a constant}$$

The constant has the dimensions of a thermal resistance (the reciprocal of thermal conductivity) and is commonly represented by θ. It may be defined as the temperature rise for unit dissipation and is commonly expressed in °C per mW.

$$T_j = T_{amb} + P_c\theta$$

θ depends on the conductivity between the collector electrode and the external surrounds and is thus a function of the physical construction of the transistor. For a given transistor θ and T_j are constant and the maximum power dissipation decreases linearly with rise in ambient temperature.

For small transistors θ depends almost entirely on the conductivity of the material between the semiconducting material and the case. For such transistors there is little reduction in θ by mounting the case on a heat sink and cooling is by convection through the air. A typical value for θ for such transistors is 0·35 °C/mW and if we take T_j as 75°C we can plot P_c against T_{amb} as shown in Fig. 8.12. This is a useful curve for it enables us to calculate the power dissipation permissible at any given value of ambient temperature. At 20°C, for example, the dissipation is 160 mW but at 40°C it has fallen to 100 mW.

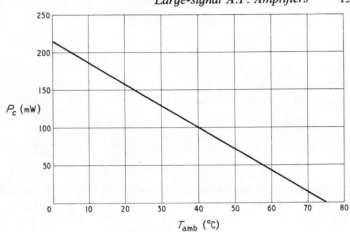

Fig. 8.12. Relationship between P_c and T_{amb} for a transistor for which $T = 75°C$ and $\theta = 0.35°C/mW$

For transistors intended for larger power outputs the collector electrode is in direct metallic contact with the case which is designed for bolting to a large block of metal such as a chassis to act as a heat sink. This is an example of conduction cooling. The thermal resistance between collector and case of such transistors is very low and the overall thermal resistance depends very largely on the properties of the heat sink.

The total or effective thermal resistance between collector junction and the outside air is made up of the following components:

1. the thermal resistance θ_1 between collector junction and transistor cases: a typical value is $1.2°C/W$;
2. the thermal resistance θ_2 across the mica washer which is generally used between case and heat sink to isolate the case electrically from the heat sink: a typical value is $0.5°C/W$;
3. the thermal resistance θ_3 of the heat sink itself. This depends on the mass and area of the sink.

The total thermal resistance is given by

$$\theta = \theta_1 + \theta_2 + \theta_3$$
$$= 1.7 + \theta_3$$

Suppose, as a numerical example, a dissipation of 6 W is required, that T_j is limited to 80°C and that the ambient temperature is 50°C. These values might be encountered in the design of the output stage of a car radio. From the expression

$$T_j = T_{amb} + \theta . P_c$$

we have

$$\theta = \frac{T_j - T_{amb}}{P_c}$$

But $\theta = 1 \cdot 7 + \theta_3$

$$\therefore \; \theta_3 = \frac{T_j - T_{amb}}{P_c} - 1 \cdot 7 = \frac{80 - 50}{6} - 1 \cdot 7$$

$$= 3 \cdot 3°\text{C/W}$$

To give a thermal resistance of this order, the heat sink can take the form of a sheet of aluminium about 40 in^2 in area and at least $\frac{1}{16}$ in thick. If the equipment has a chassis this may satisfy these requirements and the output transistor can be mounted on it. If, however, the equipment has printed wiring and requires no chassis, then a separate heat sink is necessary: the size of the sink can be reduced by providing it with cooling fins to give the required area.

Temperatures at the various junctions in the thermal circuit, e.g. at the junction between chassis and washer or between washer and transistor case can be calculated in the following manner. As already shown the thermal resistance of the heat sink is $3 \cdot 3°\text{C/mW}$. The power transmitted through it is 6 W and thus from the general expression

temp. difference $=$ power $\times$ thermal resistance

we have

temp. difference $= 6 \times 3 \cdot 3$

$$= 20°\text{C approximately}$$

The temperature of the air is 50°C, that of the chassis at the point of contact with the washer will be $50 + 20 = 70°\text{C}$. For the mica washer itself the temperature difference across it is given by $6 \times 0 \cdot 5 = 3°\text{C}$ giving the temperature at the junction between washer and transistor case as 73°C.

CHAPTER 9

D.C. and Pulse Amplifiers

The fundamental circuits of Figs. 7·7 and 7·10 for two-stage current and voltage amplifiers are also used for d.c. and pulse amplification. Voltage amplifiers often include an emitter-follower stage to give a high input resistance or a low output resistance, and negative feedback is commonly applied over three stages where one of them is an emitter follower.

D.C. AMPLIFIERS

These are amplifiers used for magnifying small voltages or currents so that slow variations in them can be more readily seen and measured. To ensure a good response at the very low frequencies of interest the amplifiers are usually direct-coupled.

D.C. amplifiers are used in electro-cardiographs, computers, pyrometers and in sensitive measuring instruments such as milli-microammeters.

Direct-coupled Cascaded Circuit

A simple d.c. amplifier could be constructed as suggested in Fig. 9.1 by using a number of transistors in cascade, each collector being direct-coupled to the base of the following transistor. Such a circuit would, however, suffer from the following serious limitations:

(1) Variations in the collector current of any of the transistors

caused by alterations in leakage current due to temperature changes give rise to a spurious output from the amplifier. Such an output is usually termed *zero drift* because it occurs with no signal input to the amplifier. The leakage-current variations in the first transistor are clearly the most important in determining the magnitude of zero drift because these are amplified by the two following transistors.

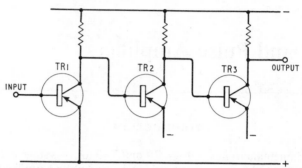

Fig. 9.1. Basic circuit for a three-stage d.c. amplifier. The performance is unsatisfactory for reasons given in the text

Such drift can be minimised by using silicon planar transistors which at normal temperatures have negligible leakage currents.

(2) Variations with temperature of the useful component βI_b of collector current can similarly give a spurious output from the amplifier. This effect can be more serious than (1) if the external base circuit of the transistors is of low resistance because there is then no means of defining I_b. Drift due to variation in I_b can be minimised by arranging for the input signal for each transistor to be supplied from a high-resistance source: the standing base current is then largely determined by the external source voltage and resistance.

(3) The gain of a common-emitter stage is determined by the value of β which is itself dependent on mean emitter current and on temperature. The gain of a simple amplifier such as that illustrated in Fig. 9.1 thus depends on the magnitude of the input signal with the result that the output–input characteristic of the amplifier is non-linear. By sacrificing sensitivity the characteristic can be made as linear as desired by use of negative feedback.

Provided all the precautions mentioned under (1) to (3) above are taken a circuit based on Fig. 9.1 can be made to give a satisfactory performance. A suitable circuit diagram is given in Fig. 9.2. The emitter bias voltages for TR2 and TR3 must be steady because changes in these would be interpreted by the transistors as input signals. Resistors, if used in each emitter circuit to give the required emitter voltage, would reduce gain by introducing negative feed-back. Individual batteries could be used but this is hardly a satis-factory arrangement because changes in battery voltage and internal resistance with age would be an embarrassment.

A solution to the problem is to use voltage-reference diodes with suitable voltage ratings in the emitter circuits. The voltage developed

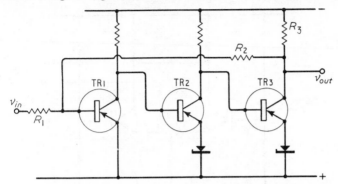

Fig. 9.2. This modification to the circuit of Fig. 9.1. is satisfactory provided silicon transistors are used

across such a diode is independent of the current in it: thus the diode behaves as a battery with constant e.m.f. and negligible internal resistance. There is thus minimal loss of signal amplification by negative feedback due to the diodes in the emitter circuits.

To give the required gain and linearity, overall negative feedback can be applied to the amplifier by a resistor R_2 connected between TR3 collector and TR1 base. The collector resistors for TR1 and TR2 must be high in value to minimise drift in TR2 and TR3, and this necessitates a fairly high supply voltage. Drift in TR1 is minimised by introducing a high-value series resistor R_1. The feedback circuit is of the type used in the basic current amplifier of Fig. 7.8 and we can therefore say immediately that the current gain from TR1 base to TR3 collector is given by R_2/R_3. Alternatively if the source of input voltage v_{in} is assumed to be of low resistance we can say that R_1 and R_2 constitute a feedback potential divider connected across the output voltage v_{out} and from page 109 the voltage gain v_{out}/v_{in} is given by R_2/R_1.

Differential Amplifier (*Long-tailed Pair*)

An alternative method of eliminating drift in direct-coupled ampli-fiers is to use two similar transistors in a balanced circuit such as that illustrated in Fig. 9.3. The transistors are connected in the common-emitter mode and base bias is provided by similar potential dividers across the supply. The transistors share a common emitter resistor, and a potentiometer enables the emitter currents to be equalised so that equal push-pull signals are received from the output terminals when an input is applied to, say, TR1. The outputs can be

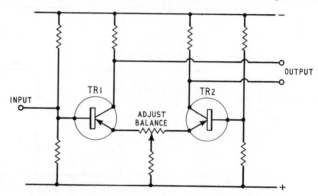

Fig. 9.3. A differential amplifier or 'long-tailed pair'

applied to a further balanced stage to increase the gain of the amplifier. The merit of this arrangement is that any change in TR1 collector current (caused by an increase in ambient temperature, for example) is accompanied by an equal increase in TR2 collector current because this transistor is of the same type as TR1 and shares the same environment. Equal increases in the collector currents cause equal increases in the voltages across the collector resistors. There is therefore no change in voltage between the output terminals as a result of the temperature change. The output of the amplifier is measured by the voltage between the output terminals and such an output can only be produced by an input applied to one of the transistors. A technique sometimes employed in differential amplifiers is to apply the input signal to TR1 base and the feedback voltage to TR2 base.

Chopper Amplifier

When drift must be kept very low or where high d.c. gain is required, a chopper amplifier is sometimes used. In this the signal to be amplified is interrupted, i.e. chopped at regular intervals to produce an alternating signal with an amplitude proportional to the d.c. signal. The alternating signal is then amplified to the required degree in a pulse amplifier, which can be RC-coupled. The output is restored to d.c. form by simple rectification or by commutation which must be synchronous with the interruption of the original signal. Chopping of the input and output signals can be achieved by mechanical means, e.g. by a vibrating reed or electronically for example by a gate circuit driven from a multivibrator.

OPERATIONAL AMPLIFIERS

General

With modern transistors it is not difficult to extend the response of a d.c. amplifier to frequencies of the order of 1 MHz. Such amplifiers are often used as operational amplifiers: these are not used for amplifying but to carry out mathematical operations such as addition or integration, the precise function being dictated by the components and circuit arrangement of a negative feedback loop. The principle of the operational amplifier can be approached in the following way.

Fig. 9.4 represents a high-gain current amplifier with a series

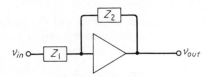

Fig. 9.4. Amplifier with negative feedback

input impedance Z_1 and a feedback impedance Z_2. From page 107 we know that the current i_b to the amplifier is given by

$$i_b = i_{in} - i_{fb}$$

where $\quad i_{in} = \dfrac{v_{in}}{Z_1} \quad$ and $\quad i_{fb} = \dfrac{v_{out}}{Z_2}$

If the gain of the amplifier is high, i_b is very small and may be neglected. The input current then becomes approximately equal to the feedback current.

$$i_{in} = i_{fb}$$

$$\therefore \frac{v_{in}}{Z_1} = \frac{v_{out}}{Z_2}$$

The effective gain of the amplifier is given by

$$\frac{v_{out}}{v_{in}} = \frac{Z_2}{Z_1}$$

which agrees with Eqn 7.4.

Now consider the circuit of Fig. 9.5 which shows two voltages v_1 and v_2 together with the output voltage all feeding current into

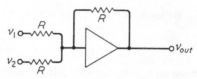

Fig. 9.5. Principle of the summing amplifier

the amplifier via equal resistors. If we again assume that $i_b = 0$ we have

$$\frac{v_1}{R} + \frac{v_2}{R} = \frac{v_{out}}{R}$$

from which

$$v_{out} = v_1 + v_2$$

i.e. the output of the amplifier is the sum of the two input voltages. Clearly this circuit can be extended to embrace any number of inputs. Moreover if v_2 is reversed in phase

$$v_{out} = v_1 - v_2$$

and the output voltage is now equal to the difference between the two input voltages.

Finally consider the circuit of Fig. 9.6. The feedback current is

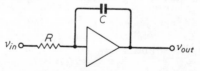

Fig. 9.6. Principle of the integrating amplifier

given by $dQ/dt = C \cdot dv_{out}/dt$ and thus we have, again assuming the amplifier input current i_b to be zero

$$\frac{v_{in}}{R} = C \cdot \frac{dv_{out}}{dt}$$

from which

$$\frac{\mathrm{d}v_{out}}{\mathrm{d}t} = \frac{v_{in}}{RC}$$

$$\therefore v_{out} = \frac{1}{RC} \int v_{in}\mathrm{d}t$$

the output voltage of the amplifier is equal to the time integral of the input voltage. This, of course, is the principle of the Miller-integrator circuit described on page 244.

Integrated-circuit Operational Amplifier

Fig. 9.7 gives the circuit diagram of a monolithic operational amplifier. It is contained in a can only $\frac{1}{3}$-in in diameter.

To minimise drift the first stage consists of a balanced pair TR2, TR3 direct-coupled to a second pair TR8, TR9. To give a high input resistance the first pair is preceded by emitter followers TR1, TR4. TR9 is followed by two common-emitter stages TR10, TR12 in cascade which drive the complementary output pair TR13, TR14. Direct coupling is used throughout to extend the response to zero frequency: this is essential if the amplifier is to be used for mathematical operations on steady potentials.

It is normally necessary for the output potential to be zero when both inputs are at zero potential. This condition can be obtained by adjustment of the collector supply voltage for TR2 and the lead from R_1 is brought out to an external connection for this purpose.

The circuit comprising R_3 to R_8 and TR5 to TR7 provides further protection against drjft. The stabilisation for the remaining part of the amplifier has something in common with that of the Tobey and Dinsdale amplifier described on page 135. R_{10}, R_{15} and R_{11} constitute a potential divider connected across the output of the amplifier which apply to TR10 emitter a fraction of the output voltage. TR10 compares this fraction with the voltage at TR9 collector in a manner similar to that in the Tobey–Dinsdale amplifier (page 135) and any errors are corrected by TR11 which adjusts its collector current to maintain the standing output voltage at $V/2$. R_{10}, R_{15} and R_{11} also provide signal-frequency feedback and the voltage gain of the amplifier from TR10 to TR14 is given approximately by $R_{11}/(R_{10}+R_{15})$, i.e. 8·2 for the resistor values shown. The connections to R_{10} are brought out to external terminals to enable this resistor to be short-circuited. By so doing the gain of the latter half of the amplifier can be increased to approximately $R_{11}/$

R_{15}, i.e. 50. With R_{10} short-circuited the overall voltage gain of the amplifier is 60,000 and the output resistance is 100 Ω.

PULSE AMPLIFIERS

General

Amplifiers such as oscilloscope Y-amplifiers and video amplifiers are required to handle signals which may have steep, almost vertical edges and also long, almost horizontal sections. A signal which has both features is a rectangular pulse and is commonly used in tests of such amplifiers which are usually known as pulse amplifiers.

The ability of an amplifier to reproduce rapid changes such as a steep edge in a signal waveform is determined by the high-frequency response of the amplifier: in fact such amplifiers must have a response good up to the frequency given by

$$f = \frac{1}{2t}$$

where t is the rise time* of the steepest edge. If the rise time is 0·1 μs the upper frequency limit is given by

$$f = \frac{1}{2 \times 0·1 \times 10^{-6}} \text{ Hz}$$

$$= 5 \text{ MHz}$$

The ability of an amplifier to reproduce very slow changes such as an almost horizontal section in a signal waveform is determined by the low-frequency response of the amplifier: the longer the section of the waveform, the better must be the low-frequency response of the amplifier to reproduce it without distortion. As a numerical example, an amplifier required to reproduce a 50-Hz square wave with less than 2 per cent sag in the horizontal sections must have a low-frequency response which is good down to at least 1 Hz. Sometimes pulse amplifiers are direct-coupled to extend the low-frequency response to zero frequency.

To summarise the above we may say that pulse amplifiers are characterised by an extremely wide frequency response: for a video amplifier, for example, the useful frequency response may extend from very low frequencies to 5·5 MHz. A statement such as this of the steady-state amplitude response of the amplifier does not, however, give complete information about its performance

---

* The rise time of a step is defined as the time taken for the signal to rise from 10 to 90 per cent of the final steady value.

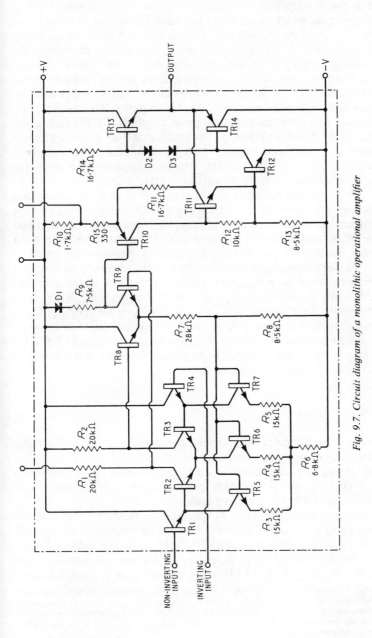

Fig. 9.7. Circuit diagram of a monolithic operational amplifier

as a pulse amplifier. In general a pulse signal contains many components which must not only undergo similar magnification but also must be maintained in their original phase relationship when distortionless amplification is required. The shunt capacitance which is inevitable in any amplifier causes the phase of high-frequency components of a complex signal such as a pulse to lag behind that of low-frequency components. Such a lag would be comparatively unimportant in an a.f. amplifier but can seriously degrade the performance of a pulse amplifier by increasing the rise time. Phase response is thus important in pulse amplifiers and for a good performance both the amplitude and the phase response must satisfy certain stringent requirements over the spectrum occupied by the signals to be amplified.

For amplifiers using simple inter-transistor coupling networks, such as are likely to be used for pulse amplification, there is a mathematical relationship between the amplitude response and the phase response so that, given one, it is possible to deduce the other. In general the flatter the frequency response the better is the phase response also and it is thus possible to ensure an adequate phase response by making the amplitude response of the amplifier sufficiently good. When a pulse amplifier is designed in this way the amplitude response must satisfy standards far more exacting than if phase response were not under consideration. For example to design a video amplifier with a passband extending to 5·5 MHz it may be necessary to make the amplitude response of the amplifier flat within 0·1 dB to say 35 MHz.

Use of Transistors

There is no difficulty in achieving bandwidths of this order using planar transistors as conventional RC-coupled amplifiers: even the transistors intended for a.f. applications have transition frequencies of about 200 MHz and transistors with f_T greater than 1 GHz, i.e. 1,000 MHz are available.

The transition frequency f_T specifies the product of the current gain and bandwidth of the transistor. As an example if a transistor with $f_T = 500$ MHz is required to give a current gain of 100, it can do this up to 5 MHz. If, on the other hand the response must be maintained up to 25 MHz the current gain is limited to 20.*

---

* In tuned amplifiers it is possible to achieve some current or voltage gain in the inter-transistor coupling circuits and this enables a transistor stage to give worthwhile gain above f_T : in pulse amplifiers, however, extra gain cannot be obtained from the resistance coupling normally used and f_T may be taken as the highest usable frequency.

Clearly in designing pulse amplifiers some means is required of dividing the product f_T between gain and bandwidth as required. This is possible by choice of load-resistor value but a better method is by use of negative feedback.

Use of Negative Feedback

Feedback is extensively used in pulse amplifiers principally to make the performance independent of the inevitable differences in characteristics between transistors of the same type and hence to make performance also independent of temperature. If a large degree of feedback is used the gain and bandwidth are largely determined by the constants of the feedback loop. Thus the performance of the amplifier can be predetermined and it is possible to manufacture a number of amplifiers all having the same performance within very close limits. This is important, for example, in colour television where the amplifiers handling the red, green and blue components of the colour signal must have identical performances.

It is not easy to apply a large degree of feedback to an amplifier without producing oscillation or a sharp peak in the frequency response curve at frequencies near the extremes of the passband. One method of avoiding such effects in a multi-stage amplifier is to employ a number of independent feedback loops each embracing only two stages (three if one is an emitter-follower).

Negative feedback reduces the gain and extends the bandwidth of an amplifier in approximately the same ratio: thus a 6 dB reduction in gain is accompanied by a doubling of bandwidth. This is illustrated in Fig. 9.8. The upper frequency-response curve is that of an amplifier without feedback and is 3 dB down at the frequency f_1 which can be taken as the limit of the passband. When feedback is applied the gain falls at all frequencies but the fall is greater at middle than at high and low frequencies and the new 3-dB loss point is now at f_2 representing an upwards extension of the passband. The low-frequency response is similarly extended but this is not illustrated in Fig. 9.8.

If the negative feedback circuit is so designed that the feedback voltage or current becomes less as frequency approaches the limits of the passband it is possible to achieve a greater extension of frequency response as suggested by the dotted line in Fig. 9.8. This, however, is usually accompanied by a deterioration in phase response, manifested by overshoot in reproduced pulses. Where a very good pulse response is essential it is advisable to adhere to the aperiodic, i.e. non-frequency-discriminating feedback represented by the solid curve. In a television receiver slight overshoot in the video amplifier may be tolerable (even advantageous by slightly exaggerating the transitions), and frequency-discriminating feedback is often used.

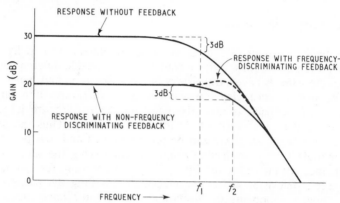

Fig. 9.8. *Effect of negative feedback on the frequency response of an amplifier*

A number of examples of typical pulse amplifier circuits will now be considered.

VOLTAGE PULSE AMPLIFIERS

Fig. 9.9 shows the circuit diagram of a voltage pulse amplifier designed to give a gain of 20 dB from a very low frequency up to approximately 6·6 MHz. Two common-emitter stages in the basic arrangement of Fig. 7.15 provide the required voltage gain and an emitter follower gives the required low output resistance. Direct coupling is used throughout and signal-frequency feedback is injected into the emitter circuit of TR1 to give the required high input resistance (approximately 10 kΩ).

Signal-frequency feedback is provided by two loops:

(a) R_9, R_{10}, R_{15} between TR3 emitter and TR1 emitter
(b) C_3, C_4 between TR2 collector and base.

Resistors R_9, R_{10} and R_{15} together with R_6 determine the gain of the amplifier over the lower middle of the passband and R_{15} is adjusted to give the required gain at 10 kHz. The same value of gain at 5·5 MHz is obtained by adjusting C_4 which, with C_3 determines the gain near the top of the passband. This rather complex feedback circuit is used to control the phase response of the amplifier at and above the upper extreme of the passband so as to eliminate any possibility of instability at these frequencies.

An essential requirement of this amplifier is that the mean output voltage should not depart significantly from zero. It is designed to operate from a supply unit giving stabilised voltages of -10 and $+14$, and a very high degree of d.c. stability in the amplifier is

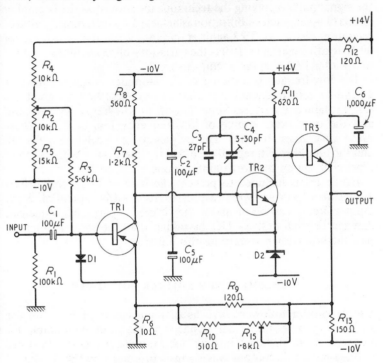

Fig. 9.9. A voltage pulse amplifier

clearly necessary. In a direct-coupled amplifier any variations in collector current of the first stage (due to changes in ambient temperature, for example) are interpreted as signals and are amplified by the following stages to give large unwanted changes in voltage at the output. To achieve high stability therefore a very low value of zero-frequency gain is wanted. For this purpose the zero-frequency gain is reduced to a very low value by the direct-coupled feedback loop R_2, R_4, R_5, R_{12} between TR3 collector and TR1 base. This loop must not, however, affect the gain of the amplifier in the passband (which extends to a very low frequency) and accordingly R_{12} is bypassed by the 1,000-μF decoupling capacitor C_6. The degree of direct-coupled feedback can be adjusted within limits by the preset potentiometer R_2 which is set to give zero voltage output from TR3 emitter. Although this feedback reduces amplification of unwanted changes in TR1 collector current, it is still possible for zero-frequency voltages in the amplifier to drift and there is need to stabilise the potential at some point in the signal path. Applying the technique mentioned at the beginning of this chapter, such stabilisation is achieved by including a voltage-reference divide in TR2 emitter circuit. With the assistance of feedback this method restricts the variations of mean voltage output to less than 0·03 V in the circuit shown.

The low-frequency response of the amplifier is not required to extend to zero frequency but must nevertheless be very good. As just explained C_6 causes some fall-off in low-frequency response and the only other source of low-frequency loss is C_1 which is made very large (100 µF). The frequency for 3 dB loss is 0·4 Hz and the amplifier can reproduce a 50-Hz square wave with less than 1 per cent sag on the horizontal sections.

Diode D1 is included to prevent the base of TR1 being driven appreciably positive with respect to its emitter potential. Such input voltages can occur and if they exceed a particular value can damage or even destroy TR1 by taking the base-emitter potential past the breakdown voltage for this junction.

COMPLEMENTARY EMITTER FOLLOWER

Simple emitter followers such as that illustrated in Fig. 7.14 are satisfactory with sinusoidal signals but may not be suitable for use with pulse signals. Consider, for example, a rectangular-wave voltage input: the output waveform is produced by the sequential charge and discharge of the shunt capacitance in the output by the emitter current of the transistor. Provided the transistor remains conductive throughout the cycle, the charge and discharge time constants are equal. The output waveform is then symmetrical and a reasonable copy of the input. If, however, the output time constant is comparable with the rise or fall times of the input, the output cannot change as rapidly as the input. It is then possible for positive-going input steps to cut off the transistor (assumed pnp). This considerably increases the output time constant by making r_o effectively infinite, leaving only R_e as the discharge path for the output capacitance. Thus for steep-sided input signals the output can have markedly dissimilar rise and fall times.

This effect can be eliminated by using a symmetrical circuit, e.g.

a complementary pair consisting of one pnp and one npn transistor feeding the same emitter resistor as suggested in the basic circuit of Fig. 9.10. Positive-going input steps, even if they cut the pnp transistor off, drive the npn transistor into conduction: similarly

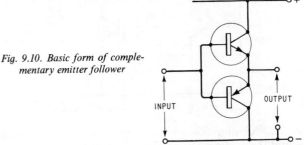

Fig. 9.10. Basic form of complementary emitter follower

negative-going input steps turn the pnp transistor on. In a practical version of this circuit it is necessary to provide the transistors with base bias: a suitable circuit is given in Fig. 8.10.

VIDEO AMPLIFIER FOR TELEVISION RECEIVER

As an example of another type of pulse amplifier we will consider the design of a video amplifier for a 625-line monochrome television receiver. This has to supply a maximum voltage swing of about 70 to the cathode of the picture tube and an output of 1·5 V (from black to white level) can be expected from the vision detector. The input resistance of the amplifier must be at least 30 kΩ to minimise damping of the detector circuit. We therefore need a voltage amplifier with a voltage gain of approximately 45, high input resistance and a response up to 5·5 MHz.

A single common-emitter stage could provide the voltage gain and frequency response but a second stage is needed to give the high input resistance and this is usually in the form of an emitter follower preceding the common-emitter stage. Thus the amplifier has the basic form shown in Fig. 9.11. To keep dissipation in the output stage at a minimum it is usual to arrange for TR2 to be almost cut off for black-level signals and to be driven into conduction by the picture signal. As the collector is direct-coupled to the tube cathode it follows that TR2 must have a positive h.t. supply and must be of npn type. TR1 gives no phase inversion and thus the diode must give an output in which increasing whiteness is portrayed by a positive excursion of voltage. This is achieved by arranging for the diode to give a negative output voltage: with

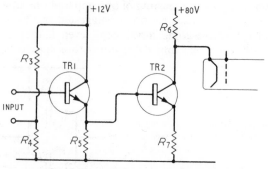

Fig. 9.11. Basic form of video amplifier

negative modulation this voltage approaches zero for peak white signals and this is the required form of operation.

The component values required in the circuit which couples TR2 to the picture tube are determined by the total shunt capacitance C_s at this point. An approximate value for C_s is 10 pF made up of contributions from tube input, TR2 output and strays. At 5·5 MHz the reactance of 10 pF is nearly 3 kΩ. If R_6 is made 3 kΩ the frequency response of the output circuit is 3 dB down at 5·5 MHz. It is preferable, however, to use a larger value for R_6 to reduce the current swing required from TR2 and to improve the gain at low and medium frequencies. A value of 3·9 kΩ is suitable. Knowing this we can determine the value of the undecoupled emitter resistor R_7. TR1 gives unity voltage gain and thus TR2 must provide the gain of 45 required. From Eqn 7.5 the gain is given approximately by R_6/R_7. Thus R_7 is equal to R_6/gain, i.e. 3,900/45 about 87 Ω.

With R_6 equal to 3·9 kΩ the 3-dB loss frequency is 4 MHz but the response can be extended to the required 5·5 MHz in two ways:

(a) by use of an inductor L_2 in series with R_6. This offsets the effect of the shunt capacitance as frequency rises. The value of L_2 can be calculated from the expression

$$L_2 = aR_6^2C_s$$

By putting $a = 0·5$ the response can be levelled at 4 MHz at the cost of approximately 7 per cent overshoot. If $R_6 = 3·9$ kΩ and $C_s = 10$ pF L_2 is 80 μH.

(b) by shunting R_7 by a capacitor C_4. This reduces the feedback due to R_7 as frequency rises. A suitable value for C_4 can be found by equating the time constants of the collector and emitter circuits of TR2. We thus have

$$R_7C_4 = R_6C_s$$

$$\therefore C_4 = \frac{R_6 C_s}{R_7}$$

$$= \text{voltage gain} \times C_s$$

$$= 45 \times 10 \text{ pF}$$

$$= 450 \text{ pF}$$

The output from the detector varies between a minimum value for peak white signals and -1.5 V at the tips of synchronising pulses. When the detector output is almost zero, the voltage across R_6 must be 70 V and across R_7 1.5 V. If, as is likely, TR2 is a silicon planar transistor, the base-emitter voltage is, say, 0.7 so that the voltage across R_5 is 2.2 V. A suitable value for R_5 is 1.5 kΩ and the current in this resistor is then 1.5 mA. If TR1 is also silicon planar the base-emitter voltage is again 0.7 and the base-earth input voltage for the amplifier is 2.9 V. As the detector provides almost zero voltage approximately 2.9 V must be obtained from R_4. The potential divider $R_3 R_4$ must be proportioned to give this voltage. It is usual to make one of the resistors preset so that correct operation of the amplifier can be obtained in spite of differences in the parameters of the transistors used.

The input resistance of TR2 is given approximately by βR_7 and if β is taken as 50, is equal to 4.35 kΩ. This is in parallel with R_5 (1.5 kΩ) giving a net external emitter resistance for TR1 of 1.1 kΩ. If β for TR1 is also taken as 50 the input resistance of the amplifier is given by $50 \times 1.1 = 55$ kΩ, nearly double the minimum value required. This is, of course, the input resistance to alternating signals but the resistance for zero-frequency signals is of the same order. We have assumed that β for both transistors is maintained at 50 up to 5.5 MHz, equivalent to $f_T = 275$ MHz. It would probably be satisfactory, however, to use transistors with $f_T = 200$ MHz because the only effect of this reduction would be to lower the input resistance of the amplifier at the upper end of the video-frequency range: a little reduction is acceptable.

To deliver the required output of 70 V TR2 requires an h.t. supply of, say, 80 V. The collector-current swing required is 18 mA and a dissipation of approximately 0.5 W is likely. TR2 must hence be capable of working under these conditions. If the receiver is intended for operation from a 12-V battery the 80-V supply can be obtained by rectifying the output of a winding on the line output transformer. TR1 can operate satisfactorily from a 12-V supply.

Fig. 9.12 gives a more complete circuit diagram of the amplifier and the preceding detector circuit used in a portable television receiver. $L_1 C_1 C_2$ constitute an i.f. filter and R_1 is the diode load

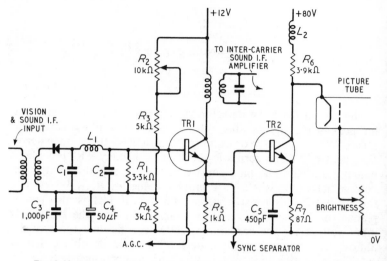

Fig. 9.12. More complete circuit diagram of the video amplifier of Fig. 9.11

resistor. The amplifier is shown direct-coupled from vision detector to picture-tube cathode. This ensures a constant black level in reproduced pictures provided that the pre-detector stages are controlled by a suitable a.g.c. system. The a.g.c. voltage must be proportional to the voltage at some point in the amplifier during back-porch periods of the video waveform: a suitable point is indicated in the diagram. Some provision is also required, in the r.f. or i.f. stages, for manual control of the input to the video amplifier to permit correct setting of contrast in reproduced pictures.

In 625-line television the sound is usually transmitted by frequency modulation and it is normal practice in receivers to amplify vision and sound together in the r.f. and i.f. stages and to abstract the sound signal after the vision detector. At this point the sound is present as frequency modulation of an inter-carrier, i.e. a carrier with a frequency equal to the difference between vision and sound carriers. TR1 may be used as an amplifier of the inter-carrier signal and the output taken from the collector circuit as indicated in Fig. 9.12.

LUMINANCE AND COLOUR-DIFFERENCE AMPLIFIERS

The problems of obtaining adequate drive voltage and bandwidth are more difficult in colour television receivers. For example the luminance amplifier (which corresponds to the video amplifier in a monochrome receiver) is required to drive all three cathodes of the colour picture tube and the capacitance at this point is further increased by the potentiometers used to adjust the relative amplitudes of the three signals. The total capacitance is of the order of 47 pF made up of 10 pF from the output of the transistor (with its heat sink), 17 pF from tube cathodes, 15 pF from potentiometers and 5 pF from strays. The reactance of 47 pF at 5·5 MHz is only 615 Ω and to generate the required 100-V amplitude across such a load requires a current which peaks up to 160 mA! It is assumed that the video stage is biased to minimum collector current for black-level signals as in the monochrome amplifier just described. To reduce the peak current to a more practical value, inductance compensation may be used and by a combination of shunt and series inductors it is feasible to use a 2·2 kΩ load resistor, which reduces the peak current to 50 mA but maintains the response up to 5 MHz. Even so a supply voltage of the order of 200 is needed and the collector current may average 30 mA, giving a total dissipation of 6 W of which half is in the transistor. Silicon transistors for such an application have been developed: they can safely dissipate the power provided an adequate heat sink is used.

When a colour receiver is used to reproduce monochrome signals the only input to the colour tube is the luminance signal just described. For colour reception, however, three colour-difference signals are derived from the luminance and chrominance information in the radiated signals: these are applied to the red, green and blue guns. Three colour-difference amplifiers are thus required: they must operate in class-A to be able to cope with the positive-going and negative-going excursions of the colour-difference signals and must be capable of supplying a total voltage excursion of 200 V up to approximately 750 kHz. Transistors capable of operation from a 280-V supply are available for such applications. Using 10-kΩ collector loads and inductance compensation the required signal amplitude and frequency response can be achieved with a mean collector current of 10 mA. The dissipation is 2·8 W and a heat sink is advisable to keep the transistors cool.

A desirable precaution against possible damage to the video output transistors by internal flash-over in the picture tube is to include resistors in the tube driving circuits. Most picture tubes have spark gaps around the electrode pins which are connected via low-inductance paths to the earthed conductive coating of the tube. This arrangement ensures that transient currents due to flash-overs are confined to the low-inductance paths.

INTEGRATED-CIRCUIT PULSE AMPLIFIER

Fig. 9.13 gives the circuit diagram of a voltage pulse-amplifier manufactured in monolithic form. It is contained in a can $\frac{1}{3}$-in in diameter.

The circuit consists of three common-emitter stages and an emitter follower all direct-coupled. Signal-frequency negative feedback is applied by the network R_2, R_3, R_4 and the resistor R_7:

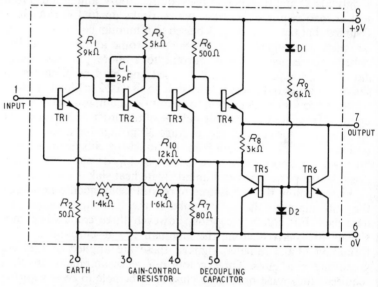

Fig. 9.13. Circuit diagram of a monolithic pulse amplifier

the connections to R_4 are brought out to pins 3 and 4 to enable gain to be adjusted by an external feedback resistor.

The method of ensuring stability of d.c. operating conditions has something in common with that used in Fig. 9.9 on page 153. The large degree of direct-coupled feedback provided by R_{10} reduces the zero-frequency gain of the amplifier to a very small value. (An external decoupling capacitor connected between pins 5 and 6 eliminates this feedback loop at signal frequencies.) The diodes D1 and D2 in conjunction with R_9, determine the base bias on TR5 and TR6. By suitable choice of diode characteristics the current in TR5 and TR6 can be kept constant in spite of variations in temperature. In this way the current in the emitter follower is stabilised as well as the d.c. conditions throughout the amplifier. So effective is this circuit that the amplifier operates satisfactorily at any temperature between $-55°C$ and $+125°C$, the variation in standing output voltage being less than 0.05%.

The upper frequency limit of the amplifier depends on the values of f_T for the transistors TR1 to TR4: it extends to 35 MHz at a gain of 200 and to 15 MHz at a gain of 400. The low-frequency response depends on the value of the external decoupling capacitor and on the values of any capacitors connected in series with the input or output. The input resistance of the amplifier is 10 kΩ and the output resistance 16 Ω. The maximum output voltage swing is 4·2 into the recommended load resistance of 1 kΩ.

CHAPTER 10

I.F. Amplifiers

The amplifiers described in earlier chapters are linear and have a frequency response beginning at a low frequency and extending for several octaves: in such amplifiers inter-transistor coupling is by a resistor or RC combination. We shall now consider linear amplifiers with much narrower frequency ranges such as i.f. amplifiers. These have bandwidths of a fraction of an octave, which are small compared with the centre frequency: in these amplifiers resonant circuits are generally employed as coupling elements. Such amplifiers usually require a level response over the passband to give uniform response to the significant sidebands of the signal and a sharp fall outside the passband to give good selectivity.

We shall now consider the design of i.f. amplifiers for sound and television receivers.

NEUTRALISED CIRCUIT

General

The frequency response required in an i.f. amplifier is often obtained by tuning one or both windings of the inter-transistor coupling transformers. The resonance frequencies, coupling coefficients and Q values must be chosen to give the required shape and extent of frequency response. The effective Q value of an inductor depends on its physical construction and also on any damping due to the components connected to it. Transistor damping is sometimes made heavy and to obtain a desired Q value, the designer must know the

162

input and output resistances of the transistors at the operating frequencies. Transistor manufacturers usually quote these parameters for frequencies of interest. Over the narrow frequency range of an i.f. amplifier the input and output resistances of a transistor may usually be taken as constant. Thus a bipolar transistor used

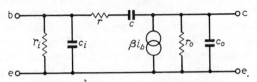

Fig. 10.1. Equivalent circuit of a bipolar transistor used
as an r.f. or i.f. amplifier

for i.f. amplification may be represented by the simple equivalent circuit of Fig. 10.1.

If the maximum possible gain is required, the inter-transistor coupling circuits must match the output resistance of each transistor to the input resistance of the next. In addition the coupling circuits must, of course, provide the required frequency response. If, however, attempts are made to realise the maximum gain, internal feedback in the transistors will almost certainly cause oscillation. This feedback occurs via an internal path of resistance and capacitance (represented by r and c in Fig. 10.1) between the collector and the base of each transistor. To avoid such instability the internal feedback can be neutralised by external components: circuits for achieving this are described later. Alternatively stability can be achieved by limiting the gain: this too is considered later.

Calculation of Maximum Gain

For calculating the maximum gain obtainable from an i.f. transistor stage the circuit of Fig. 10.2 may be used. This shows a transformer coupling TR1 to TR2 but for simplicity all tuning and

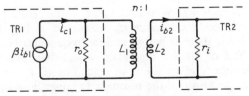

Fig. 10.2. Circuit for calculating the maximum gain of
a transistor i.f. amplifier

neutralising components are omitted. We know from page 123 that the optimum turns ratio for the transformer is $n:1$ where

$$n = \sqrt{\frac{r_o}{r_i}}$$

When matching is perfect the input resistance r_i appears effectively as a resistance r_o across the primary winding: the alternating component i_{c1} of TR1 collector current divides equally between the output resistance r_o and the transformer winding. The current in the primary is thus $i_{c1}/2$ and this gives rise to a current $ni_{c1}/2$ at the secondary winding. This is the input current i_{b2} of TR2 and thus we have

$$i_{b2} = \frac{i_{c1}}{2}\sqrt{\left(\frac{r_o}{r_i}\right)}$$

Now $i_{c1} = \beta i_{b1}$

$$\therefore \frac{i_{b2}}{i_{b1}} = \frac{\beta}{2}\sqrt{\left(\frac{r_o}{r_i}\right)} \tag{10.1}$$

This gives the maximum current gain possible from TR1 and its coupling circuit. If the input resistances of TR1 and TR2 are equal, Eqn 10.1 also gives the maximum voltage gain available from TR1.

Eqn 10.1 is sometimes quoted in terms of the mutual conductance instead of β. We know from page 56 that $\beta = g_m r_i$ and thus we have

$$\frac{i_{b2}}{i_{b1}} = \frac{g_m r_i}{2}\sqrt{\left(\frac{r_o}{r_i}\right)}$$

$$= \frac{1}{2} g_m \sqrt{(r_i r_o)} \tag{10.2}$$

Insertion Loss

In a practical circuit L_1 is usually tuned to the centre frequency by a parallel-connected capacitor C_1 as shown in Fig. 10.3. In the resulting circuit there is a loss at low frequencies because of the shunting effect of the low reactance of L_1: there is also a loss at high frequencies due to the low reactance of C_1. Ideally there should be no loss around the resonance frequency of $L_1 C_1$ because these frequencies constitute the required passband. However the dynamic resistance of $L_1 C_1$ would need to be infinite to give zero loss in the passband. In practice the finite value of dynamic resistance inevitably introduces a loss, known as the *insertion loss*, the magnitude of which can be calculated in the following way.

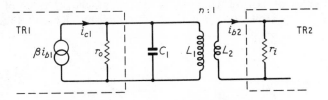

Fig. 10.3. Basic circuit for single-tuned amplifier

At the centre frequency L_1C_1 presents two resistances to the collector circuit of TR1. One is the input resistance of TR2 which, due to the action of the transformer, is effectively equal to r_o in parallel with L_1. The second resistance is the dynamic resistance R_d of the resonant circuit L_1C_1. This is also effectively in parallel with L_1 as shown in Fig. 10.4. The output resistance r_o of TR1,

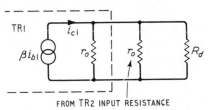

FROM TR2 INPUT RESISTANCE

Fig. 10.4. Circuit for calculating insertion loss

and the input resistance of TR2 are together equivalent to a resistance of $r_o/2$ and thus the equivalent circuit has the form shown in Fig. 10.5.

Ideally all of the output current i_{c1} should flow in $r_o/2$ but some

Fig. 10.5. Simplification of the circuit of Fig. 10.4

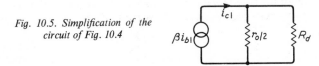

is lost in R_d and this accounts for the insertion loss. The current which flows in $r_o/2$ is given by

$$\text{current in } r_o/2 = i_{c1} \cdot \frac{R_d}{r_o/2 + R_d}$$

The insertion loss is thus given by

$$\text{insertion loss} = 20 \log_{10} \frac{R_d}{r_o/2 + R_d} \qquad (10.3)$$

R_d is given by $Q_u L_1 \omega$ where Q_u is the (undamped) Q value of the inductor L_1. However when L_1 is connected to TR1 and TR2 as

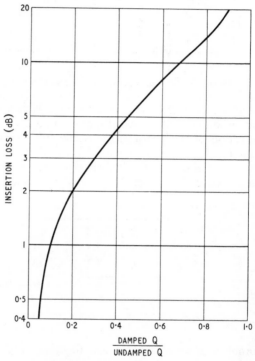

Fig. 10.6. Dependence of insertion loss on the ratio of damped to undamped Q values

shown in Fig. 10.3, damping due to the output resistance of TR1 and the input resistance of TR2 reduce the effective Q value to Q_d. In fact the dynamic resistance R_d is reduced to

$$\frac{R_d r_o/2}{R_d + r_o/2}$$

so that we can say

$$\frac{Q_d}{Q_u} = \frac{1}{R_d} \cdot \frac{R_d r_o/2}{R_d + r_o/2}$$

$$= \frac{r_o/2}{R_d + r_o/2} \qquad (10.4)$$

We can write Eqn 10.3 in the form

$$\text{insertion loss} = 20 \log_{10}\left(1 - \frac{r_o/2}{r_o/2 + R_d}\right)$$

$$= 20 \log_{10}\left(1 - \frac{Q_d}{Q_u}\right) \quad (10.5)$$

This expression is plotted in Fig. 10.6. This shows that if the Q of the tuned circuit is halved by the connection of the two transistors, the insertion loss is 6 dB. If the damping is increased to reduce the effective Q to one-third the undamped value, the insertion loss falls to 3·5 dB. Light damping, giving a small reduction in Q value, gives a great insertion loss. Thus the more selective the amplifier is made, the greater is the insertion loss and the lower the gain: this paradox seems to contradict familiar gain-bandwidth considerations.

Determination of Inductance and Capacitance

If the transformer is designed to match r_o to r_i, the resistance effectively damping $L_1 C_1$ is $r_o/2$. This reduces the dynamic resistance of $L_1 C_1$ from its undamped value of $Q_u L_1 \omega$ to $Q_d L_1 \omega$ where $Q_d L_1 \omega$ is the parallel resistance of $r_o/2$ and $Q_u L_1 \omega$. Thus we have

$$Q_d L_1 \omega = \frac{\dfrac{r_o}{2} Q_u L_1 \omega}{\dfrac{r_o}{2} + Q_u L_1 \omega}$$

from which

$$L_1 = \frac{r_o(Q_u - Q_d)}{2\omega Q_d Q_u} \quad (10.6)$$

Now

$$C_1 = \frac{1}{\omega^2 L_1}$$

$$\therefore C_1 = \frac{2Q_d Q_u}{\omega r_o(Q_u - Q_d)} \quad (10.7)$$

For very heavy damping $Q_d \ll Q_u$ and we have

$$L_1 \approx \frac{r_o}{2\omega Q_d} \quad (10.8)$$

$$C_1 \approx \frac{2Q_d}{\omega r_o} \quad (10.9)$$

Neutralisation

In the following simplified account of feedback via the internal collector-base capacitance c_{re}, effects due to frequency discrimination in the external base circuit are ignored.

At resonance the external collector circuit is effectively resistive and the signal voltage at the collector is in antiphase with the collector current. The current fed back to the base via the reactive path provided by c_{re} is therefore in quadrature with the external base input current.

Above the resonance frequency the external collector circuit is effectively capacitive and the phase of the collector voltage, with respect to the collector current, lags the antiphase condition by 90°. The current in c_{re} is in antiphase with the external base current, giving negative feedback and reduced gain from the transistor.

Below the resonance frequency the external collector circuit is effectively inductive. The phase of the collector voltage relative to the collector current leads the antiphase condition by 90°. The current in c_{re} is in phase with the external base current giving positive feedback and increased gain from the transistor. If the feedback current exceeds the external input the circuit will oscillate. Even when the feedback is insufficient to cause oscillation, the change from positive to negative feedback at the resonance frequency of the collector circuit can cause asymmetry in the frequency response curve. This can be avoided by reducing the gain of the amplifier as explained on page 178 or, if the full gain is wanted, by the use of external components which neutralise the effects of c_{re}.

One method of eliminating internal feedback is illustrated in Fig. 10.7. From a point where the signal voltage is in antiphase

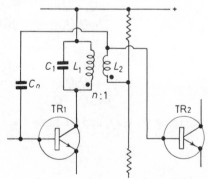

Fig. 10.7. One circuit for neutralising a transistor tuned amplifier

with that at the collector, a capacitor C_n is connected to the base of the transistor. In the circuit illustrated the antiphase signal is obtained from the secondary winding which feeds the following transistor: the two dots mark the ends of the windings where the signals are in phase. If the transformer has a turns ratio of $n:1$ as shown, C_n should be made equal to nc_{re}. Thus if $c_{re} = 1$ pF and $n = 7$, C_n should be 7 pF.

An alternative circuit in which the antiphase signal is obtained by tapping the inductor of the collector circuit is shown in Fig. 10.8. Here the matching of TR1 to the following transistor is

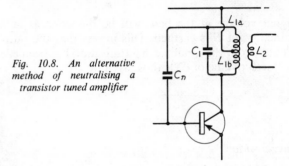

Fig. 10.8. An alternative method of neutralising a transistor tuned amplifier

achieved by correct choice of turns ratio between L_{1b} and L_2, and the required value of C_n is determined by the turns ratio between L_{1a} and L_{1b}.

Alternatively the capacitive branch of the tuned circuit can be tapped. To do this C_1 is replaced by two capacitors C_{1a} and C_{1b} connected in series. If the junction of the two capacitors is earthed, as shown in Fig. 10.9(a), antiphase signals are generated across C_{1a}. It is the signal at the collector end of L_1 which causes unwanted

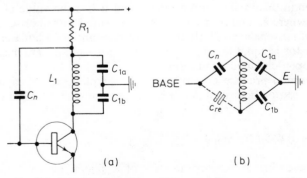

Fig. 10.9 A third method of neutralising a transistor tuned amplifier

feedback via c_{re} and neutralisation can be effected by connecting C_n between the top end of L_1 and the transistor base. To avoid short-circuiting C_{1a}, the supply is introduced into L_1 via a resistor R_1 which should be large compared with the reactance of C_{1a}. R_1 and C_{1a} may appear to be decoupling components but C_{1a} is too small for this purpose.

The four capacitors c_{re}, C_n, C_{1a} and C_{1b} form a bridge circuit as shown in Fig. 10.9(b). Provided

$$\frac{C_n}{c_{re}} = \frac{C_{1a}}{C_{1b}}$$

the signal voltage at the base will be the same as at E, namely zero since point E is earthed. This means that the output signals do not appear at the input, i.e. neutralisation has been effected.

The correct value of C_n is thus given by

$$C_n = c_{re} \cdot \frac{C_{1a}}{C_{1b}} \qquad (10.10)$$

Unilateralisation

The internal feedback between collector and base of a transistor is via a path which includes resistance and capacitance. In early i.f. amplifiers it was customary to employ a network of R_n and C_n in series in the external neutralising circuit. By using the correct values for R_n and C_n it is possible to offset the internal feedback completely so that the transistor becomes truly a one-way device in which the input circuit is completely divorced from the output circuit. This type of perfect neutralising is known as *unilateralisation*. For a unilateralised transistor the input resistance is unaffected by conditions in the output circuit and it has the value of h_{ie} or $1/g_{ie}$, where h_{ie} and g_{ie} are the values measured with short-circuited output terminals. Similarly the output resistance of a unilateralised transistor is equal to $1/h_{oe}$ or $1/g_{oe}$ where h_{oe} and g_{oe} are measured with open-circuited input terminals.

Common practice now is to use the simple forms of capacitive neutralisation described above or to dispense with it altogether.

Design of 465-kHz Stage

The following are the parameters of a silicon planar transistor intended for use at 465 kHz. They apply at 1-mA mean collector current, a value likely in a battery-operated portable receiver where current economy is important.

$$g_{ie} = 300 \text{ μmho} \qquad g_{oe} = 4 \text{ μmho} \qquad g_m = 35 \text{ mA/V}$$

$$c_{ie} = 25 \text{ pF} \qquad c_{oe} = 1.4 \text{ pF} \qquad c_{re} = 1 \text{ pF}$$

$$f_T = 300 \text{ MHz}$$

If the transistor is used in a neutralised or unilateralised circuit $r_i = 1/g_{ie}$, i.e. 3·3 kΩ and $r_o = 1/g_{oe}$, i.e. 250 kΩ. If we assume that the following transistor also has an input resistance of 3·3 kΩ we can calculate the maximum gain from Eqn 10.2 thus

$$\text{maximum gain} = \frac{g_m}{2}\sqrt{(r_i r_o)}$$

Substituting $g_m = 35$ mA/V, $r_i = 3.3$ kΩ and $r_o = 250$ kΩ we have

$$\text{maximum gain} = 500 \text{ approximately}$$

In decibels

$$\text{maximum gain} = 20 \log_{10} 500 = 54 \text{ dB}$$

For an amplifying stage of the basic form shown in Fig. 10.7 we can calculate the required values of L_1 and C_1 in the following way. A suitable value for the passband of an a.m. medium- and long-wave receiver is 7 kHz. The required value of Q_d is thus given by

$$Q_d = \frac{\text{centre frequency}}{\text{passband}} = \frac{465}{7} = 66.5$$

For a practical miniature inductor the maximum value of un-damped Q is 130 and for this the insertion loss is given by Eqn 10.5 thus

$$\text{insertion loss} = 20 \log_{10}(1 - Q_d/Q_u) = 20 \log_{10}(1 - 66.5/130)$$

$$= 6 \text{ dB approximately}$$

The maximum value of gain is thus 250 (48 dB).

The value of L_1 can be calculated from Eqn 10.6

$$L_1 = \frac{r_o(Q_u - Q_d)}{2\omega Q_u Q_d}$$

Substituting $r_o = 250$ kΩ, $Q_u = 130$, $Q_d = 66.5$ and $\omega = 2\pi \times 465$ kHz we have

$$L_1 = 314 \text{ μH}$$

From Eqn 10.7 we have

$$C_1 = \frac{2Q_u Q_d}{\omega r_o (Q_u - Q_d)}$$

Substituting the numerical values given above

$$C_1 = 372 \text{ pF}$$

For maximum gain the turns ratio of the transformer is given by

$$\sqrt{\left(\frac{250,000}{3,300}\right)} : 1 = 8 \cdot 7 : 1$$

If the neutralising circuit of Fig. 10.7 is used the required value of C_n is given by $8 \cdot 7 c_{re}$, i.e. approximately 9 pF.

To achieve the required centre frequency and passband, the tuning capacitance C_1 in the collector circuit must be 372 pF as calculated above. It is not, however, necessary to use a physical capacitor of this value. It is often more convenient to use a standard capacitor of say 200 pF (together with an inductor which resonates with this capacitance at 465 kHz) and to connect the collector to a tapping point on the inductor, as shown in Fig. 10.10, such that

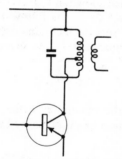

Fig. 10.10. Simplified circuit for an i.f. stage using a standard value of tuning capacitance

the effective capacitance in the collector circuit is 372 pF. The tapping point should be at $\sqrt{\dfrac{200}{372}}$, i.e. 0·73 of the total number of turns from the earthy end and the number of secondary turns should be $1/8\cdot7$ of the turns included in the collector circuit.

Final I.F. Stage

In the circuits discussed above the output resistance of the transistor amplifier was assumed matched to the input resistance of the following stage. This is, of course, the condition for securing maximum gain from the transistor but it results in limited output power which is almost certainly insufficient for good operation of a diode detector. This limitation results from the high value of the effective load resistance of such a matched stage: this is equal to the output resistance of the stage and is probably many times the optimum load resistance. As a result inefficient use is made of the available collector-current swing. For example in the numerical example just considered the transistor is likely to have a 9-V supply and can at best give a peak collector voltage swing of 8 V (1 V being allowed for emitter bias). The effective load resistance for a matched stage is 250 kΩ and the peak current swing is thus $8/250 = 0.03$ mA, only 1/30th the available current swing. Thus the power output is only 1/30th that which could be delivered into the optimum load.

The final stage of an i.f. amplifier is normally designed to enable the transistor to give maximum power output. The effective load resistance for the transistor should therefore be equal to the optimum load resistance which, from page 122, is given by

$$\text{optimum load resistance} = \frac{\text{available collector-voltage swing}}{\text{available collector-current swing}}.$$

As just explained in a battery-operated receiver the available collector-voltage swing may be 8 V and the collector current swing 1 mA giving an optimum load resistance of 8 kΩ.

The circuit coupling the final i.f. stage to the detector must hence present the transistor with the required 8-kΩ resistance. As the aim is to deliver as much power as possible to the detector the insertion loss due to the tuned circuit must not be too great: nevertheless some contribution towards the selectivity of the receiver is also desirable. An undamped Q of 130 which is effectively reduced to 50 is a reasonable compromise giving an insertion loss of 4 dB. Let us now consider the design of a suitable form of tuned transformer for such a final i.f. stage.

Assume that a tuning capacitance of 250 pF is chosen and that Q_u for the inductor is 130. The undamped dynamic resistance of the tuned circuit is given by

$$\frac{Q_u}{\omega C_1} = \frac{130}{6.284 \times 465 \times 10^3 \times 250 \times 10^{-12}} \Omega$$

$$= 178 \text{ k}\Omega$$

This must be damped by the detector circuit to give an effective Q value of 50, for which the damped dynamic resistance is 68.4 kΩ.

The connection of the detector circuit must reduce the dynamic resistance from 178 to 68·4 kΩ: the damping due to the detector circuit must thus be equivalent to connecting a resistance R across the tuned circuit where

$$68\cdot4 = \frac{178R}{178+R}$$

This gives

$$R = \frac{178 \times 68\cdot4}{178 - 68\cdot4}$$

$$= 111 \text{ k}\Omega$$

For an a.m. receiver a normal value for the diode load is 5 kΩ and if the detector were 100 per cent efficient the input resistance would be one-half of this. However for the small inputs likely in a transistor receiver with a limited supply voltage, the diode detector efficiency is not high and the input resistance is likely to be approximately 3 kΩ. The turns ratio of primary to secondary winding is thus given by $n:1$ where

$$n = \sqrt{\frac{111}{3}} = 6\cdot07$$

The collector can be connected to a tapping point on the primary winding. The damped dynamic resistance of the tuned circuit is 68·4 kΩ and to give the required 8-kΩ load the tapping point should give a turns ratio of

$$\sqrt{\frac{68\cdot4}{8}} = 2\cdot925:1$$

The turns ratio from collector circuit to diode input is thus 6·07/2·925, i.e. 2·076:1 and from this we can calculate the gain. We know that the voltage gain from base to collector is given by $g_m R_c$ where g_m is 35 mA/V and R_c is 8 kΩ. This gives a gain of 280. From base to detector input the voltage gain is hence 280/2·076, i.e. 135. The insertion loss due to the tuned circuit has been allowed for in this calculation.

Limitations of the Neutralised Circuit

Although the neutralised circuit can give high gain it has a number of limitations. The provision of neutralising components is a difficulty because their value must be such that stability is assured in spite of the inevitable spreads in transistor parameters.

Secondly the performance of the circuit is likely to be affected by a.g.c. action. Because of the tight coupling between tuned circuits and transistors, variations in the input and output capacitance of the transistors can alter the centre frequency of the transformers. Variations in input and output resistances can affect the passband.

UNNEUTRALISED CIRCUIT

General

For the reasons just given it is common practice to dispense with neutralising and to ensure stability in i.f. amplifiers by limiting the gain by suitable choice of source and load resistance. The load resistance can be given the optimum value thus enabling the transistor to make full use of the available voltage and current swings and so deliver maximum power output. Although gain is limited it is still possible using this design technique to obtain adequate i.f. gain for a television receiver from three cascaded transistors.

Stability Factor

Appendix C shows that a transistor with similar tuned circuits tuned to the same frequency in collector and base circuits will oscillate if

$$\omega c_{re} g_m R_b R_c = 2 \tag{10.11}$$

where R_b and R_c are the resistances effectively in parallel with the base and collector circuits.

For the transistor used in previous numerical examples $g_m = 35$ mA/V and $c_{re} = 1$ pF. If the input and output resistances are matched to the source and load we have at 465 kHz

$$\omega c_{re} g_m R_b R_c = 6\cdot284 \times 465 \times 10^3 \times 1 \times 10^{-12} \times 35 \times 10^{-3}$$
$$\times 3,300 \times 250,000$$
$$= 84\cdot4$$

This is considerably greater than the critical value of 2 and thus oscillation is inevitable unless neutralisation is used. Even when the gain is reduced by using an 8-kΩ load the critical value is still exceeded and neutralisation is still required. These calculations confirm earlier statements that neutralisation is essential if the full gain of a transistor is required.

To be certain of stability the left-hand side of Eqn 10.11 must be less than two and the factor by which the expression must be multiplied to give two is known as the *stability factor*. For a stability factor of 4

$$\omega c_{re} g_m R_b R_c = 0.5$$

Stability factors between 2 and 8 are used in designing amplifiers without neutralising, the chosen value depending on likely spreads in g_m and c_{re} and on such considerations as whether single or double-tuned circuits are used. Such amplifiers are said to be designed to have *stability-limited gain*. We shall now consider the design of an i.f. amplifier of this type for an f.m. sound receiver. The principle can also be applied, of course, to the design of a 465-kHz amplifier.

10.7-MHz I.F. Stage

For f.m. receivers the standard intermediate frequency is 10.7 MHz and for transmissions with a maximum rated deviation of ± 75 kHz a bandwidth (to the -3 dB points) of 200 kHz is needed to accept the significant sidebands of the transmission. For adequate selectivity the response is required to be -40 dB at 600 kHz bandwidth. To give such a response it is common to use three double-tuned transformers in the i.f. amplifier. Each transformer is thus required to be -1 dB at 200 kHz and -13.3 dB at 600 kHz bandwidth. The flattest response is obtained from a pair of identical coils when these are critically coupled and for this degree of coupling, universal selectivity curves show that the response is -1 dB when

$$\frac{Q \times \text{bandwidth}}{\text{centre frequency}} = 1$$

This gives the required value of damped Q as

$$Q_d = \frac{\text{centre frequency}}{\text{bandwidth}} = \frac{10.7}{0.2}$$

$$= 53$$

Consider now one of the i.f. transistors situated between two of the critically coupled bandpass filters. The parameters of the transistor are not likely to differ significantly from those quoted above for the 465-kHz stage. We thus have, for a stability factor of 4:

$$\omega c_{re} g_m R_b R_c = 0.5$$

$$\therefore R_b R_c = \frac{0.5}{\omega c_{re} g_m}$$

Substituting for ω, c_{re} and g_m

$$R_b R_c = 2.174 \times 10^5 \qquad (10.12)$$

This product is considerably less than that of the input and output resistances of the transistor and it follows that the source and load resistances cannot match the transistor resistances. We can thus choose a value for R_c and a convenient value to select is 8 kΩ for this enables the transistor to deliver maximum power output. Substituting for R_c in Eqn 10.12 gives R_b as 27 Ω.

The problem now is to design the bandpass filters to present the transistor with these values of source and load resistance. An important point is that the dynamic resistance of a tuned circuit is halved when a similar circuit is critically coupled to it. If therefore we design the tuned circuits to give the required source and load resistances in the absence of coupling to their companion tuned circuits, the resistances decrease and stability increases when the couplings are present. This means that instability is not possible even if the tuned circuits are grossly mistuned (as during alignment), mistuning being equivalent to reduction of coupling coefficient. Decrease of load resistance means that the transistor cannot deliver its maximum power but this is unlikely to cause difficulty in early i.f. stages. For the final stage the design should aim at maintaining the optimum load.

The calculated source resistance is small compared with the input resistance of the transistor and the chosen load resistance is small compared with the output resistance. Thus the transistor does not significantly damp the preceding or following tuned circuit and to give the required passband the coils can be designed to have un-damped Q values of 53.

If the whole of the tuned circuit is connected directly in the collector circuit R_d must be 8 kΩ. As Q is 53 we can calculate the required tuning capacitance as follows:

$$R_d = \frac{Q_d}{\omega C}$$

$$\therefore C = \frac{Q_d}{\omega R_d} = \frac{53}{6.284 \times 10.7 \times 10^6 \times 8 \times 10^3} \text{ F}$$

$$= 100 \text{ pF approximately.}$$

The inductance can similarly be calculated

$$R_d = L\omega Q_d$$

$$\therefore L = \frac{R_d}{\omega Q_d} = \frac{8 \times 10^3}{6 \cdot 284 \times 10 \cdot 7 \times 10^6 \times 53} \text{ H}$$

$$= 2 \cdot 245 \text{ }\mu\text{H}$$

A tuned circuit with these values of L and C is also used to feed the base of the transistor. The circuits have an (uncoupled) dynamic resistance of 8 kΩ and to obtain the 27 Ω effective source resistance the secondary winding must be tapped, the position of the tapping point being given by

$$\sqrt{\left(\frac{27}{8,000}\right)} = 0 \cdot 0337$$

from the earthy end of the tuned circuit.

CALCULATION OF GAIN

To calculate the gain of the 10·7-MHz stage suppose that the following stage has a similar transistor fed from a similar tapping point on the secondary winding of the double-tuned transformer. The effective resistance of the primary circuit is 4 kΩ so that the voltage gain between base and collector is given by

$$g_m R_c = 35 \times 10^{-3} \times 4 \times 10^3 = 140$$

The voltage across the secondary winding is equal to that across the primary and the gain to the tapping point is thus $140 \times 0 \cdot 0337 = 9 \cdot 13$ (18 dB).

TRANSISTORS WITH LOW c_{re}

This gain is much smaller than for the same transistor at 465 kHz primarily because of the greater feedback via c_{re} at 10·7 MHz. Gain can be improved if c_{re} can be reduced and transistors with c_{re} of 0·15 pF are now available. For these the product $R_b R_c$ (for a stability factor of 0·5) is nearly seven times that calculated above and it is thus possible to let $R_c = 8$ kΩ as before and R_b 190 Ω. This enables the secondary tapping point to be raised to

$$\left(\frac{190}{8,000}\right) = 0 \cdot 154$$

and the voltage gain then becomes $140 \times 0 \cdot 15 = 21$ (26 dB).

USE OF CAPACITANCE TAPPING

Inductors of the order of 2 µH such as are required in 10·7-MHz
i.f. amplifiers are unlikely to have more than approximately 10 turns
and it is therefore difficult to obtain desired positions of tapping
points with precision. An alternative solution is to tap the capacitive
branch of the tuned circuit instead. For example to obtain a tapping
point 0·154 along an inductor as required in the last numerical
example we can proceed thus. If the capacitor of 100 pF is replaced
by a series combination of two capacitors C_a and C_b then to keep
the tuning capacitance at 100 pF we have

$$\frac{C_a C_b}{C_a + C_b} = 100 \text{ pF} \qquad (10.13)$$

The 'tapped down' output is of course taken from the larger of the
two capacitors. If this is C_b we have

$$\frac{100}{C_b} = 0\cdot154$$

This gives

$$C_b = 650 \text{ pF}$$

Substituting in Eqn 10.13 above

$$\frac{650 C_a}{650 + C_a} = 100$$

from which

$$C_a = 120 \text{ pF}$$

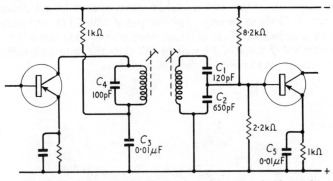

Fig. 10.11. *A stage of a 10·7MHz i.f. amplifier using double-tuned
transformers and capacitive tapping*

A typical 10·7-MHz i.f. stage using capacitance tapping is shown in Fig. 10.11.

TELEVISION I.F. AMPLIFIERS

Television i.f. amplifiers require bandwidths wider than any so far described. For example for 625-line television a bandwidth of approximately 5·5 MHz is required to accommodate the vision sidebands but the sound signal is also normally included and this extends the bandwidth to 6 MHz. The intermediate frequency (corresponding to the vision carrier frequency) is 39·5 MHz and the required passband is from 33·5 to 39·5 MHz, the centre frequency being 36·5 MHz.

A single-tuned circuit with such values of centre frequency and bandwidth needs a working Q value given by

$$Q_d = \frac{\text{centre frequency}}{\text{bandwidth}} = \frac{36\cdot5}{6} = 6\cdot1$$

much lower than any so far considered. Undamped Q values are very large compared with this and thus insertion losses are likely to be very low. This is important in the tuned circuit following the output stage because of the desirability of getting maximum signal into the detector.

In a mains-driven receiver the final i.f. stage might be designed to take a mean collector current of 10 mA from a 20-V supply. The optimum load is then approximately 2 kΩ and this is also the approximate input resistance of the vision detector (a load resistance of 4 kΩ is assumed). The coupling circuit can thus take the form of a 1:1 transformer with the detector circuit providing the damping. The output resistance of the final i.f. stage, even at 10-mA collector current, is likely to be very large compared with 2 kΩ. The damped dynamic resistance of the collector load is thus 2 kΩ: from this and the damped Q value of 6·1 we can calculate the tuning capacitance and inductance required in the following way:

$$R_d = \frac{Q_d}{\omega C}$$

$$\therefore C = \frac{Q_d}{\omega R_d}$$

$$= \frac{6\cdot1}{6\cdot284 \times 36\cdot5 \times 10^6 \times 2 \times 10^3} \text{ F}$$

$$= 13\cdot3 \text{ pF}$$

a small capacitance but a feasible value to use in view of the extremely low values of output capacitance of silicon planar transistors. The inductance value can be calculated thus

$$L = \frac{R_d}{\omega Q_d}$$

$$= \frac{2 \times 10^3}{6 \cdot 284 \times 36 \cdot 5 \times 10^6 \times 6 \cdot 1} \text{ H}$$

$$= 1 \cdot 43 \text{ }\mu\text{H}$$

As the transformer has unity turns ratio the current gain of the final i.f. stage is equal to β, e.g. 50. To avoid instability in an un-neutralised stage (as this is assumed to be) the effective value of the source resistance should be calculated as indicated earlier in this chapter and the input circuit designed to provide this.

The earlier stages of the i.f. amplifier could be designed in a similar manner to that just described, using the input resistance of the following stage to provide the required damping. It would be difficult however with a succession of synchronously tuned circuits each with a Q of 6, even with additional rejector tuned circuits, to provide the steep fall-off outside the passband that is required to prevent inter-ference from sound or vision signals on neighbouring channels. This difficulty can be resolved by using higher Q values such as 20 or 30 in the earlier i.f. stages and by tuning them to frequencies such as 35 MHz and 38 MHz displaced from the centre frequency. By suit-able choice of Q values and resonance frequencies, this method of stagger tuning can give the required shape of response curve and good selectivity, although it is still necessary to use additional rejector circuits. The design of these earlier stages can follow the procedure indicated above for 'stability limited gain' stages, the external collector resistance being so chosen that the transistor can supply the required amplitude of signal, the external base resistance being made low enough to provide an adequate margin of protection against instability.

Three transistors can supply all the gain required in a television i.f. amplifier and a block diagram of a typical arrangement is given in Fig. 10.12. Only the first stage is automatic gain controlled and the input circuit is so designed that variations in input capacitance and input resistance have negligible effect on the shape of the frequency response curve. The input tuned circuit is commonly one-half of a bandpass pair, the other half being in the tuner (which contains the r.f. and f.c. stages). So that the channel selection controls are conveniently placed, the tuner is often at some distance from the

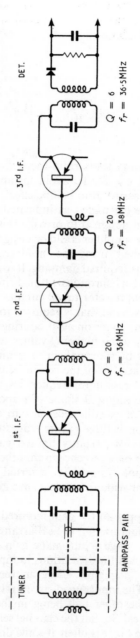

Fig. 10.12. Basic form of stagger-tuned television i.f. amplifier

remainder of the receiver and is coupled to it by coaxial cable, the capacitance of which provides some of the shunt-capacitance coupling between the two halves of the bandpass filter.

AUTOMATIC GAIN CONTROL

Most receivers use some form of automatic gain control to minimise the effects of ionospheric or man-made* fading and to ensure that all signals, irrespective of their input amplitude, are reproduced at substantially constant amplitude. A.g.c. is achieved by controlling the gain of pre-detector, usually i.f., stages by a voltage (or current) derived from the signal at the detector output or at a post-detector point. Some means is therefore required of adjusting the gain of a transistor amplifier by a control voltage.

The gain of a transistor falls at low collector currents and at low

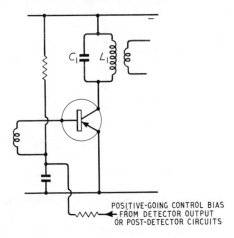

Fig. 10.13. An example of reverse a.g.c.

collector-emitter voltages and there are two corresponding ways of achieving gain control. The first is by applying the control voltage to the base as a reverse bias: this is known as *reverse control* and an example, using a pnp transistor and positive-going control bias, is given in Fig. 10.13. An npn transistor would, of course, require a negative-going bias for reverse control. For both types of transistor the control bias increases when a strong signal is received thus

* e.g. aircraft flutter in television reception.

biasing back the transistor and reducing the gain. An unfortunate feature of this type of control is that the signal-handling capacity of a transistor is reduced by reverse bias but the circuit has the advantage that collector current is reduced when strong signals are received: this is important in battery-operated receivers where current economy is desirable.

In the second method, known as *forward control*, the gain of the transistor is reduced by increasing the forward bias, thus increasing collector current. An essential feature of the circuit, illustrated in Fig. 10.14, is the decoupling circuit R_2C_2: as the collector current

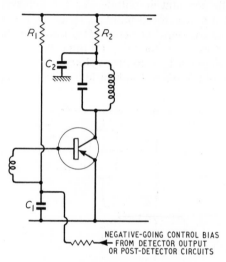

NEGATIVE-GOING CONTROL BIAS
◄—FROM DETECTOR OUTPUT
OR POST-DETECTOR CIRCUITS

Fig. 10.14. An example of forward a.g.c.

increases, the voltage drop across R_2 is increased thus reducing the collector-emitter voltage and forcing the operating point to move into the knee of the I_c–V_c characteristics where the characteristics are more crowded and the g_m therefore lower. For this type of control a pnp transistor requires a negative-going bias and an npn a positive-going bias. Forward control has the advantage of increasing the signal-handling capacity of the transistor when this is needed for large-amplitude signals. Not all transistors are suitable for forward control and it is important to select a type which has been specifically designed for use in this type of circuit. By increasing the collector current of a suitable transistor from 4 to 13 mA, it is possible to reduce the gain by more than 40 dB.

Both methods of a.g.c. are in common use, sometimes in the same receiver. Both forms of control reduce the power output of the controlled stage. This is of little significance in early i.f. stages but it is not usual to apply a.g.c. to the final i.f. stage because this is required to supply appreciable power to the detector.

In Fig. 10.11 the source of base input signal (i.e. the secondary winding of the i.f. transformer) and the emitter decoupling capacitor are both returned to the positive terminal of the power supply. This minimises the impedance of the external base-emitter circuit which is essential for maximum performance. The primary winding of the i.f. transformer must be returned to the negative terminal of the supply to provide the necessary collector bias but should be returned to the positive terminal of the supply in order to minimise the impedance of the external collector-emitter circuit which is also essential for best results. This difficulty is usually overcome by connecting a low-reactance capacitor C_3 (the collector decoupling capacitor) between the primary winding and the emitter (or the positive terminal of the supply which is in turn connected to the emitter via the low-reactance emitter decoupling capacitor). If the collector decoupling capacitor is omitted, the signal output current of the transistor must flow through the impedance of the supply in addition to the primary winding of the transformer. In this way signal voltages are developed across the battery and these, if impressed upon other stages of the amplifier or receiver of which this stage is part, can distort the shape of the frequency response curve or even cause instability. To avoid this the collector decoupling capacitor C_3 is included to short-circuit the supply at signal frequencies.

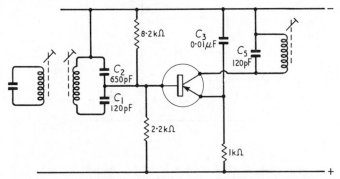

Fig. 10.15. *A modified version of the circuit of Fig. 10.11. which eliminates the need for collector decoupling components*

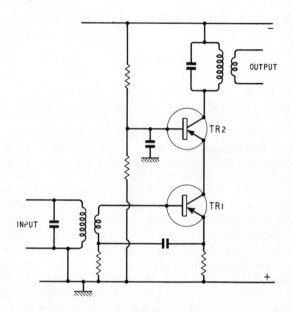

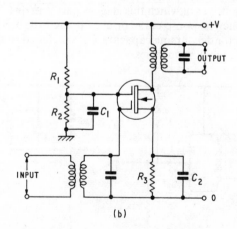

(b)

Fig. 10.16. A cascode i.f. or r.f. amplifier
using (a) two bipolar transistors and (b) a
dual gate i.g.f.e.t.

However, by a simple alteration to the circuit such a capacitor becomes unnecessary. If the source of base input signal (i.e. the secondary winding) and the emitter decoupling capacitor are both returned to the negative terminal of the supply, as shown in Fig. 10.15, then the impedance of the external base-emitter circuit and of the external collector-emitter circuit are minimised simultaneously and no additional decoupling capacitor is necessary.

CASCODE CIRCUIT

So far we have mentioned two ways in which instability via the internal base-collector capacitance of a transistor can be avoided: one is by neutralisation and the other by limiting the gain of the amplifier. A third method is by the use of the cascode circuit. In essentials this consists of a common-emitter stage feeding into a common-base stage and a typical circuit is shown at (a) in Fig. 10.16. The common-emitter stage has such a low value of collector load (the input resistance of the common-base stage) that the voltage gain is limited to a value which makes instability via the collector-base capacitance impossible. The base of the common-base stage is effectively earthed at signal frequencies so avoiding instability in this stage. The cascode has an input resistance and mutual conductance equal to those of a common-emitter stage and may be regarded in fact as a stable common-emitter amplifier.

The cascode circuit is also used with f.e.t.s. A single common-source f.e.t. stage is almost impossible to use as an i.f. or r.f. amplifier because of instability caused by feedback via the drain-gate capacitance and high input resistance. A pair of f.e.t.s may, however, be successfully used as a cascode amplifier in an arrangement similar to that using bipolar transistors, i.e. a common-source stage feeding into a common-gate stage. There is, however, no need to use two separate transistors: a single dual-gate i.g.f.e.t. can be used as shown in Fig. 10.16(b) and there is no need for the drain-source connection between the transistors to be brought out. This circuit has much in common with that of a tetrode or screened-pentode i.f. amplifier, the upper gate behaving as a screen: in fact the dual-gate f.e.t. is sometimes known as a tetrode f.e.t.

INTEGRATED-CIRCUIT I.F. AMPLIFIER

Fig. 10.17 gives the circuit diagram of a monolithic i.f. amplifier. TR1 and TR2 form a cascode amplifier and TR3, the a.g.c. transistor,

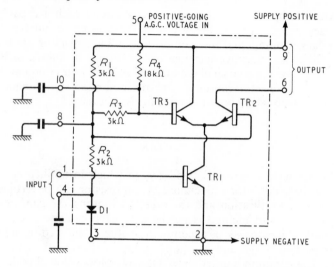

Fig. 10.17. Circuit diagram of a monolithic i.f. or r.f. amplifier

is in parallel with TR2 and its output load. TR2 and TR3 are biased from the same point on the potential divider $R_1 R_2$ and TR1 is biased from the diode D_1 which ensures constancy of mean current in TR1.

When TR3 is forward biased it shunts the output load and reduces the output-signal level. The shunting effect becomes more significant as the positive a.g.c. voltage on TR3 base is increased. This is an effective method of controlling gain and has the merit of leaving the input resistance and input capacitance of the amplifier substantially constant.

An integrated circuit of this type can be housed in a can about $\frac{1}{3}$-in in diameter and is capable of giving 25 dB gain up to 60 MHz at 6-MHz bandwidth.

I.C.s are likely to be used increasingly for i.f. amplification. They can provide high and stable gain and, for use in f.m. receivers, amplitude limiting as well. Such i.c.s in conjunction with ceramic filters (which need no tuning adjustment) will probably be the form of future i.f. amplifiers.

CHAPTER 11

Sinusoidal Oscillators

INTRODUCTION

Sinusoidal oscillators have numerous applications: for example they are used in radio transmitters, receivers, carrier telephony equipment and test instruments. This chapter describes the principles of a number of common types of transistor oscillator.

Sinusoidal oscillators have two main sections, a frequency-determining section and a maintaining section. The frequency-determining section commonly consists of an LC or RC network. The maintaining section is a transistor amplifier (with its power supply) which must have sufficient gain to offset the attenuation of the frequency-determining section and must also introduce the degree of phase shift required by this section.

LC OSCILLATORS

In general, transistor LC oscillators are of the same basic types as valve oscillators and their circuits can be deduced by analogy with the valve circuits. It may, however, be desirable to modify the circuit or the component values to suit the input and output impedances of transistors where these differ from those of a valve.

When a signal is momentarily induced in an LC circuit, it gives rise to an oscillation which is at the resonance frequency of the circuit and dies away exponentially due to dissipation in the resistance of the circuit. To maintain the amplitude of the oscillation the loss of power must be made good. This can be achieved by connecting the LC circuit to a source of power such as an amplifier. This source can be regarded as having a negative input or output resistance which neutralises the positive resistance of the tuned circuit. In the Hartley, Colpitts and Reinartz oscillators the negative resistance is achieved by positive feedback and it is characteristic of these types of oscillator that at least three connections are necessary between the frequency-determining and amplifier sections to obtain this feedback.

Hartley Oscillator

Fig. 11.1 gives the circuit diagram of a Hartley oscillator. The steady component of the collector current is stabilised by the potential-divider method, and, except for the components required for stabilisation, the circuit is similar to that of a valve Hartley oscillator. The frequency-determining section L_1C_1 is connected between collector (anode) and base (grid). The emitter (cathode) is

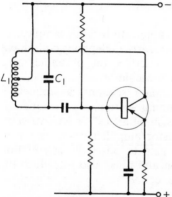

Fig. 11.1. One possible circuit for a transistor Hartley oscillator

effectively connected to the tapping point on the inductor because the impedance of the power supply can be assumed to be negligible at the frequency of oscillation. The three essential connections providing positive feedback are thus to the two ends of the coil and to the tapping point.

The circuit illustrated is a series-fed type in which the collector current flows through part of the inductor. There is an alternative, shunt-fed type in which the transistor has a resistor as collector load and the collector is connected to the inductor via a capacitor.

Colpitts Oscillator

Fig. 11.2 gives the circuit diagram of a Colpitts oscillator, also stabilised by the potential-divider method. The three connections providing positive feedback are to the two ends of the coil and to the junction of C_1 and C_2.

In this circuit C_1 is in parallel with the output capacitance of the transistor and C_2 is in parallel with the input capacitance. Thus the transistor capacitances have an effect on the oscillation frequency and if the capacitances change (due, for example, to a change in supply voltage or ambient temperature) there is a corresponding change in the operating frequency. This effect can be

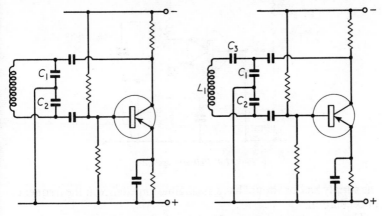

Fig. 11.2. Shunt-fed Colpitts oscillator   *Fig. 11.3. Clapp or Gouriet oscillator providing good frequency stabilisation*

minimised, and good frequency stability achieved, by making C_1 and C_2 large compared with the transistor capacitances. A further capacitor C_3 is then included, in series with L_1 (Fig. 11.3). C_3 is made small compared with C_1 and C_2: the resonance frequency is then substantially that of L_1 and C_3. Very good frequency stability can be achieved in this circuit known as the Clapp or Gouriet oscillator.

Reinartz Oscillator

The Reinartz oscillator (Fig. 11.4) is of interest because it, or a derivative of it, is frequently used in transistor receivers. Positive feedback is achieved here by coupling the collector circuit to the emitter circuit via the mutual inductance between L_1 and L_2. Both inductors are also coupled to the frequency-determining circuit L_3C_3. This oscillator is also stabilised by the potential divider-method but in this circuit the lower arm of the potential divider must be decoupled by a low-reactance capacitor as shown to ensure that the signals developed across L_2 are applied directly between base and emitter. The frequency-determining section of the Reinartz oscillator appears at first sight to have four connections (two from L_1 and two from L_2) but the connections to the positive and negative terminals of the supply are effectively common because

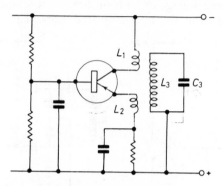

Fig. 11.4. Reinartz oscillator

the supply has, or should have, negligible impedance at the frequency of oscillation.

This circuit can be simplified by omitting the components L_3 and C_3 and connecting a tuning capacitor across L_1 or L_2 but the form illustrated here is often preferred in receivers because it permits the moving vanes of the tuning capacitor to be earthed: in addition the loose coupling between $L_3 C_3$ and the transistor helps to improve frequency stability.

PHASE-SHIFT OSCILLATORS

General

The three types of oscillator described above all employ a single common-emitter transistor amplifier. One of the characteristics of this type of amplifier is that the output (collector) signal is a magnified but phase-inverted reproduction of the input (base) signal. Thus to obtain the positive feedback essential for oscillation the frequency-determining circuit must (at the operating frequency) also introduce a phase inversion between the connections to the collector and to the base; moreover, the attenuation between these connections must be less than the gain of the amplifier. Provided these two conditions are satisfied the form of the frequency-determining circuit does not matter. LC circuits were used in the oscillators described above but RC circuits can also be used as described below.

RC Oscillators

A single RC circuit such as that illustrated in Fig. 11.5 introduces a phase shift which increases as frequency falls and approaches a limiting value of 90° as frequency approaches zero. At such low frequencies, however, the attenuation introduced by the circuit is very great. Two such sections in cascade could produce 180° phase

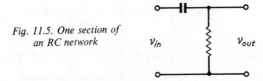

Fig. 11.5. One section of an RC network

shift (which, for a sinusoidal signal, is equivalent to phase inversion) but the attenuation would be enormous. To give 180° phase shift with reasonably small attenuation, a network of at least three sections is necessary. For use with transistor amplifiers the network should preferably require a high-impedance source (the collector circuit) and a low-impedance termination (the base circuit). A suitable form of network is $R_1 C_1 R_2 C_2 R_3 C_3$ in Fig. 11.6 which gives the circuit diagram of a complete RC oscillator.

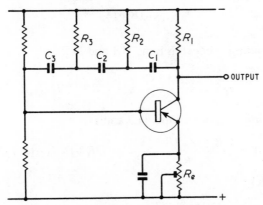

Fig. 11.6. Simple phase-shift oscillator suitable for single frequency operation

The frequency of oscillation and the attenuation of the network can be determined as follows. Normally all the frequency-determining resistors are equal. Thus

$$R = R_1 = R_2 = R_3$$

Similarly the frequency-determining capacitors are usually equal,

$$\therefore\ C = C_1 = C_2 = C_3$$

To simplify the analysis we shall assume the network to be supplied from a constant-current source and that the network is terminated in a short circuit. Thus the circuit has the form shown in Fig. 11.7

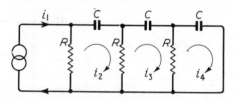

Fig. 11.7. A three-section RC network

in which i_1 is the signal current from the collector circuit and i_4 is the signal current into the base circuit. From Kirchhoff's law we have

$$i_1R = i_2(2R+jX)-i_3R \tag{11.1}$$

$$i_2R = i_3(2R+jX)-i_4R \tag{11.2}$$

$$i_3R = i_4(R+jX) \tag{11.3}$$

where $X = -1/\omega C$.

Eliminating i_2 between Eqns 11.1 and 11.2

$$i_1R = \frac{(2R+jX)^2}{R}\cdot i_3 - (2R+jX)i_4 - i_3R \tag{11.4}$$

Eliminating i_3 between Eqns 11.3 and 11.4

$$i_1R = \frac{(2R+jX)^2(R+jX)}{R^2}\cdot i_4 - (2R+jX)i_4 - (R+jX)i_4$$

from which

$$\frac{i_4}{i_1} = \frac{R^3}{R^3+6jXR^2-5RX^2-jX^3} \tag{11.5}$$

For $180°$ phase shift between i_4 and i_1, the terms in j vanish and we have

$$6jXR^2 = jX^3$$

giving

$$X = \sqrt{6}R \qquad (11.6)$$

$$\therefore -\frac{1}{\omega C} = \pm\sqrt{6}R$$

and

$$\omega = \frac{1}{\sqrt{6}RC}$$

The frequency of oscillation is thus given by

$$f = \frac{1}{2\pi\sqrt{6}RC} \qquad (11.7)$$

At this frequency we have, from Eqn 11.5

$$\frac{i_4}{i_1} = \frac{R^3}{R^3 - 5RX^2}$$

But, from Eqn 11.6

$$X^2 = 6R^2$$

$$\therefore \frac{i_4}{i_1} = \frac{R^3}{R^3 - 30R^3}$$

$$= -\frac{1}{29} \qquad (11.8)$$

Thus the transistor must give a current gain of at least 29 to achieve oscillation. This is the required ratio of base current to current in R_1 (Fig. 11.6) and some of the output current is shunted through the parallel paths presented by C_1R_2, $C_1C_2R_3$, etc. A ratio of i_4 to i_1 of 29 thus represents a high order of gain and it is desirable to select a transistor with a high value of β to give oscillation. On the other hand, to obtain a sinusoidal output the transistor must not oscillate too strongly and the gain should be adjusted to give only a small amplitude of oscillation. One way of controlling the gain is by adjustment of the value of emitter resistor R_e as shown in Fig. 11.6. Reduction in R_e increases I_e, hence increases g_m and gain. If excessive gain is used the oscillation amplitude will be checked by the collector current reaching zero or the collector voltage reaching the base voltage and the output wave-form then contains some linear sections and is distorted.

A convenient value for the frequency-determining resistors is 4·7 kΩ because this permits a collector current of 1 or 2 mA with a low supply voltage. To give oscillation at say 1 kHz the capacitance required can be calculated from Eqn 11.7 thus

$$C = \frac{1}{2\pi\sqrt{6}fR}$$

$$= \frac{1}{6\cdot284 \times 2\cdot45 \times 10^3 \times 4\cdot7 \times 10^3} \text{ F}$$

$$= 0\cdot015 \text{ μF approximately.}$$

In practice the frequency is unlikely to be precisely 1 kHz because the transistor output resistance is not infinite and its input resistance is not zero as assumed in the analysis.

This form of oscillator can readily be adapted for variable frequency oscillation. The three resistors can be ganged to form a control with a frequency range of 10:1 and the capacitors can be adjusted in decade steps to extend the range further. The attenuation of the network is always 29 at the frequency of oscillation and thus the output amplitude remains substantially constant.

Wien-bridge Oscillator

In the RC oscillator positive feedback was obtained because, in effect, both frequency-determining network and amplifier introduced a phase inversion of the signal passing through it. Alternatively, of course, oscillation could be obtained by arranging for

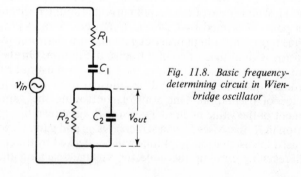

Fig. 11.8. Basic frequency-determining circuit in Wien-bridge oscillator

both network and amplifier to introduce zero phase shift. This is the principle of the Wien-bridge oscillator.

The frequency-determining network employed has the form illustrated in Fig. 11.8. The upper arm consists of resistance and capacitance in series, the lower arm of an equal resistance and an equal capacitance in parallel. The network is supplied from a constant-voltage source and is terminated in an infinite impedance: the phase shift and attenuation introduced by the network can then be calculated as follows. The impedance of the parallel RC network is $RjX/(R+jX)$ and thus

$$\frac{v_{out}}{v_{in}} = \frac{\dfrac{RjX}{R+jX}}{R+jX+\dfrac{RjX}{R+jX}}$$

where $X = -1/\omega C$. Simplifying we have

$$\frac{v_{out}}{v_{in}} = \frac{RjX}{(R+jX)^2 + RjX}$$

$$= \frac{RjX}{R^2 - X^2 + 3RjX}$$

Rationalising

$$\frac{v_{out}}{v_{in}} = \frac{RjX(R^2 - X^2) + 3R^2X^2}{(R^2 - X^2)^2 + (3RX)^2} \tag{11.9}$$

When there is zero phase shift the terms in j vanish and we have

$$R^2 = X^2$$

giving the frequency of zero phase shift as

$$\omega = \frac{1}{RC}$$

$$\therefore f = \frac{1}{2\pi RC}$$

Substituting $R = X$ in Eqn 11.9 to find the attenuation at this frequency we have

$$\frac{v_{out}}{v_{in}} = \frac{3R^2X^2}{9R^2X^2} = \frac{1}{3}$$

The maintaining amplifier thus requires a gain just exceeding 3 to sustain oscillation. Such a low voltage gain could readily be obtained from a single transistor but it is difficult in such a simple amplifier to obtain the required phase shift at the same time as the desired terminating resistances. Usually, therefore, a two- or even three-stage amplifier is used and the gain is reduced to the required value by negative feedback.

Wien-bridge oscillators are frequently used for a.f. testing purposes. The oscillation frequency can be readily adjusted if a two-gang variable capacitor is used for C_1 and C_2 (giving a 10:1 change in frequency), the resistors R_1 and R_2 being switched in decade steps to give different frequency ranges. By this means it is possible to design an oscillator with a frequency range from say 30 Hz to 30 kHz.

The capacitors used for C_1 and C_2 must be variable and are unlikely therefore to have a maximum capacitance greater than 500 pF. For this capacitance value the values of R_1 and R_2 required to give oscillation at 30 Hz can be calculated by substitution in the above expression. We have

$$R = \frac{1}{2\pi f C}$$

$$= \frac{1}{6 \cdot 284 \times 30 \times 500 \times 10^{-12}} \, \Omega$$

$$= 10 \text{ M}\Omega \text{ approximately.}$$

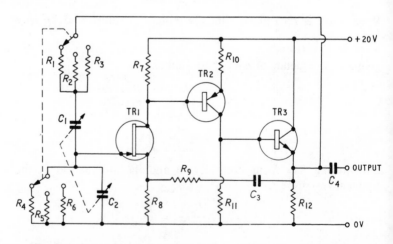

Fig. 11.9. Circuit diagram for a Wien-bridge oscillator with range switching

Fig. 11.9 gives the circuit diagram of one possible form for an oscillator of this type. The amplifier has three transistors including one common-source stage, one common-emitter stage and an emitter follower output stage. The first two stages give phase inversion and the third does not, so that the required phase characteristic is obtained. The emitter follower gives a low output resistance for feeding the frequency-determining network and the output terminals of the oscillator. As already shown the resistance in the lower arm of the frequency-determining network must be several megohms to give oscillation at low audio frequencies. To avoid resistive shunting of this arm (which would raise the oscillation frequency) the input resistance of the amplifier must be very high and an f.e.t. is the obvious choice for TR1. TR1 and TR2 can, of course, give voltage gain far greater than the value of 3 required to sustain oscillation and it is essential to reduce the gain to this value otherwise oscillation amplitude grows until limited by transistor current cut-off and the required sinusoidal waveform is not obtained. Negative feedback is used to reduce the gain to the required value and to stabilise the amplitude of the output signal.

Direct coupling is used throughout the amplifier. This gives simplicity of design and ensures a good low-frequency response. The low value of zero-frequency gain also ensures good stability of mean transistor currents.

The output amplitude of the oscillator is stabilised by making the negative feedback dependent on it. This could be achieved by using a resistor with a negative resistance-temperature coefficient (e.g. a thermistor) for R_9. R_9 is fed from the amplifier output via C_3 and the dissipation in it and hence its resistance is entirely dependent on the oscillation amplitude. Alternatively a resistor with a positive resistance-temperature coefficient (e.g. a lamp with a metal filament) could be used for R_8, R_9 being a non-temperature-dependent resistor. In both circuits any tendency for the output amplitude to increase, increases negative feedback, decreases amplifier gain and thus checks the rise in amplitude.

CHAPTER 12

Detectors, Frequency Changers and Receivers

INTRODUCTION

Some of the circuits used in sound and television receivers, such as a.f. amplifiers, video amplifiers and i.f. amplifiers are described in other chapters. In this chapter we shall be concerned with other aspects of receivers such as detectors, r.f. amplifiers and frequency changers. Finally we shall consider the circuits of complete sound receivers.

A.M. DETECTORS

The purpose of the detector in a receiver is to derive from a modulated carrier a substantially-undistorted copy of the modulation waveform. Frequently an a.m. detector also has to produce an output

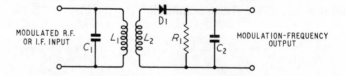

Fig. 12.1. Basic form of diode detector circuit

signal, proportional to the carrier amplitude, which can be used for automatic gain control and in a transistor receiver appreciable power is necessary for a.g.c.

Although a transistor may be used in a circuit similar to that of

200

an anode-bend detector, it is more usual to use a semiconductor diode as an a.m. detector. The basic circuit is shown in Fig. 12.1.

The diode D1 conducts during positive-going half-cycles of modulated r.f. input and charges C_2 to a voltage approximately equal to the peak value of the r.f. signal across L_2. During negative-going half-cycles D1 is cut off and C_2 discharges through the diode load resistor R_1. Thus a voltage is developed across R_1C_2 which rises and falls in sympathy with the modulation waveform but never reverses in polarity: in Fig. 12.1 the voltage at the diode output terminal is always positive.

Choice of Time Constant

The time constant R_1C_2 is important: if it is too small C_2 discharges too rapidly and the detector output is unnecessarily low; if it is too large C_2 cannot discharge rapidly enough to follow the most rapid variations in the modulation waveform and distortion results. The optimum value of the time constant is the largest which just fails to cause distortion: this value can be calculated in the following way.

If the detector is perfect, an unmodulated carrier of amplitude v_{in} gives a steady voltage of v_{in} across the load. In practice the output voltage has a ripple made up of a succession of exponential rises and falls, the rises being governed by the time constant r_dC_2 (r_d is the diode forward resistance) and the falls by time constant R_1C_2. Normally R_1 is large compared with r_d and the falls are slower than the rises. The ripple is usually eliminated by a simple RC filter in the detector output circuit.

If the carrier input is sinusoidally modulated to a depth m, a sinusoidal signal of mv_{in} amplitude is superimposed on the steady voltage v_{in} across the detector load and this too is composed of exponential rises and falls. To avoid a distorted output the exponential falls must be rapid enough to follow the sinusoidal form. If the modulation waveform has the equation

$$v = mv_{in} \sin \omega t$$

the rate of change is given by

$$\frac{dv}{dt} = mv_{in}\omega \cos \omega t$$

The greatest value this can have occurs when $\cos \omega t = 1$ and is equal to $mv_{in}\omega$: the sine wave has this slope where it crosses the datum line.

The voltage at the datum line is v_{in} and the slope of the exponential fall which commences at this point is given by v_{in}/R_1C_2 as shown on page 237. Equating slopes we have

$$mv_{in}\omega = \frac{v_{in}}{R_1C_2}$$

$$\therefore R_1C_2 = \frac{1}{m\omega} = \frac{1}{2\pi fm}$$

The time constant is thus inversely proportional to frequency and we must choose a value which will not cause distortion at the highest modulating frequency: it will then automatically be suitable for lower frequencies.

In a 625-line television receiver the upper limit of the video band is approximately 5 MHz and signals at this end of the band can be 100 per cent modulated. Thus $f = 5 \times 10^6$, m = 1 and we have

$$R_1C_2 = \frac{1}{6{\cdot}284 \times 5 \times 10^6} \text{ s}$$

$$= 0{\cdot}032 \text{ } \mu\text{s}$$

In high-quality sound broadcasting the upper frequency limit is 15 kHz and if signals at this extreme were 100 per cent modulated the required time constant would be given by

$$R_1C_2 = \frac{1}{6{\cdot}284 \times 15 \times 10^3} \text{ s}$$

$$= 10{\cdot}6 \text{ } \mu\text{s}$$

However in sound broadcasting the high modulating frequencies are harmonics and have amplitudes small compared with those of lower-frequency fundamental signals. Moreover in a.m. receivers the i.f. amplifier usually restricts the upper frequency limit to 3 or 4 kHz by sideband cutting. Thus it is quite permissible to increase R_1C_2 in an a.m. sound receiver to 50 µs or more.

When the time constant has been decided we can consider the individual values of R_1 and C_2. R_1 must not be too small otherwise C_2 cannot charge up to the peak value of the input signal and the output of the detector is unnecessarily low: to minimise this loss R_1 should be large compared with r_d the forward resistance of the diode. On the other hand R_1 must not be too large otherwise the shunting effect of the following stages becomes serious: moreover in television receivers large values of R_1 can lead to impossibly-low values for C_2. A compromise value of R_1 commonly adopted in sound receivers is 5 kΩ but in television receivers slightly lower values such as 3·3 kΩ are sometimes used.

If R_1 for a television receiver is 3·3 kΩ we can calculate C_2 from the time constant thus

$$C_2 = \frac{0\cdot032 \times 10^{-6}}{3\cdot3 \times 10^3} \text{ F}$$

$$= 10 \text{ pF}$$

If R_1 for a sound receiver is 5 kΩ we can calculate C_2 from the time constant thus

$$C_2 = \frac{50 \times 10^{-6}}{5 \times 10^3} \text{ F}$$

$$= 0\cdot01 \text{ μF}$$

Input Resistance

If the detector is 100 per cent efficient C_2 is charged up to the peak value $v_{in\,(pk)}$ of the input signal. Thus the output power delivered by the detector to the load resistor is given by $v_{in\,(pk)}{}^2/R_1$. If the input resistance of the detector is r_i the input power supplied to the detector by the signal source is given by $v_{in\,(rms)}{}^2/r_i$. If no power is lost in the diode we can equate these two expressions thus

$$\frac{v_{in\,(rms)}{}^2}{r_i} = \frac{v_{in\,(pk)}{}^2}{R_1}$$

$$\therefore r_i = \frac{v_{in\,(rms)}{}^2}{v_{in\,(pk)}{}^2} \cdot R_1$$

For a sinusoidal signal

$$v_{in\,(rms)} = v_{in\,(pk)}/\sqrt{2}$$

and thus

$$r_i = \frac{R_1}{2}$$

In practice some power is inevitably lost in the forward resistance of the diode and the output voltage is less than expected: r_i then is larger than $R_1/2$ and a value of 3 kΩ is commonly assumed when R_1 is 5 kΩ.

A.G.C. Provision

The voltage required for a.g.c. may be positive or negative depending whether pnp or npn transistors are to be controlled and on whether forward or reverse control is required. Either polarity can be obtained by suitable design of the detector circuit. For example Fig. 12.2 gives the complete circuit of a detector for a

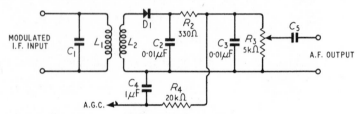

Fig. 12.2. *Complete circuit diagram for a diode detector in a sound receiver*

sound receiver. R_2C_3 attenuates i.f. ripple in the detector output and the diode load R_3 functions also as the volume control. R_4C_4 is an a.f. filter which minimises a.f. content in the a.g.c. voltage which in this circuit is positive-going. By reversing the diode a negative-going control voltage can be obtained.

F.M. DETECTORS

The function of an f.m. detector is to derive from a frequency-modulated signal a substantially-undistorted copy of the modulating waveform impressed on the signal. There are many types of f.m. detector but only two are in common use in f.m. receivers: these are the Foster-Seeley discriminator and the ratio detector. The method of operation of these detectors is complex and only a brief summary of it is given below: for a more complete description the reader is referred to other books.*

Foster-Seeley Discriminator

This discriminator contains two diode detectors so arranged that their outputs are connected in series opposition. The diodes are fed from a double-tuned transformer, the primary and secondary

* For example, King, Gordon J. *FM Radio Servicing Handbook*, 2nd edition, Newnes–Butterworths.

windings of which are resonant at the centre frequency of the passband to be covered. An essential feature of the circuit is that a
fraction of the primary voltage is fed to the centre point of the
secondary winding. In Fig. 12.3 this is achieved by a connection
between the centre point of L_2 and a tapping point on L_1 but the
secondary connection could be to the junction of two equal capacitors across L_2 (they could together constitute the tuning capacitance) and the primary connection could be to an inductor closely
coupled to L_1 or to a capacitive potential divider across L_1 (formed
by two capacitors which may also provide the tuning capacitance).

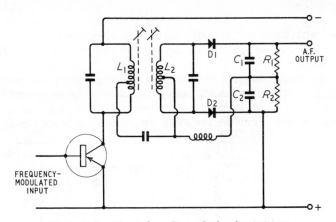

Fig. 12.3. One circuit for a Foster-Seeley discriminator

For signals at the centre frequency, diodes D1 and D2 receive
equal inputs and the voltages generated across R_1 and R_2 are equal,
giving zero resultant voltage across $(R_1 + R_2)$. The effect of the
interconnection between primary and secondary windings is that
for signals displaced from the centre value one diode receives a
bigger input than the other. The voltages across R_1 and R_2 are
then no longer equal and there is a net output across $(R_1 + R_2)$,
the polarity depending on the direction of the frequency displacement and the magnitude depending on the extent of the displacement. If, therefore, a frequency-modulated signal is applied to the
discriminator, a copy of the modulation waveform is generated
across $(R_1 + R_2)$.

The Foster-Seeley discriminator gives zero output at the centre
frequency and at this frequency the output is independent of the
magnitude of the signal input to the detector. At other frequencies
the output of the discriminator is proportional both to frequency
displacement and to signal input. Ideally the output of an f.m.

detector should be proportional to the frequency displacement but independent of the signal input. If this can be achieved the full advantages of frequency modulation are realised and the receiver is to a large extent immune from interference due to a.m. signals and from the distortion due to multi-path reception. The Foster-Seeley discriminator thus has poor ability to reject a.m. signals and is normally preceded by separate limiter stages in order to obtain a satisfactory performance.

Ratio Detector

The ratio detector has much better a.m. rejection than a Foster-Seeley circuit and nearly all commercial receivers employ a ratio detector.

The circuit diagram of one form of ratio detector is given in Fig. 12.4. It has two diodes and a double-tuned transformer with a primary-secondary connection similar to that employed in the Foster-Seeley circuit but the diodes are connected in a series-aiding

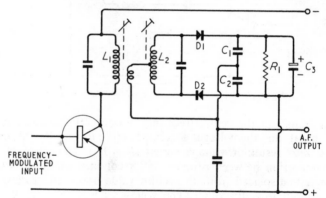

Fig. 12.4. One circuit for a ratio detector

arrangement and supply a common load resistor. This resistor has a low value to give the heavy damping of the secondary tuned circuit on which the limiting properties of the detector depend. The load resistor is bypassed by a high-value capacitor giving a load time constant of 0·1 s. The diodes conduct continuously when a signal is applied to the detector and give a voltage across the load circuit proportional to the carrier input: the maximum value of this voltage gives an indication of the correct tuning point. The inputs to the two diodes vary with frequency displacement (as in the Foster-

Seeley circuit) and the voltages generated across the reservoir capacitors C_1 and C_2 vary also although the total voltage across $(C_1 + C_2)$ is independent of input frequency, being stabilised at a value proportional to carrier input amplitude by the long time constant R_1C_3. One end of the capacitor C_3 is earthed usually and the a.f. output is taken from the junction of C_1 and C_2.

Practical ratio detector circuits commonly employ additional resistors in series with the diodes: by choosing suitable values for these resistors, the a.m. rejection of the detector can be significantly improved.

MIXERS

In a superheterodyne receiver the carrier and sidebands constituting the received signal are in effect translated in frequency to give a new signal with a carrier at the intermediate frequency. This is achieved in a mixer stage in which the received signal is combined with the output of a local oscillator, thereby producing a resultant signal at the sum or difference of the received carrier and oscillator frequencies.

In general, there are two basic principles which can be applied to produce an output with a frequency equal to the difference between the frequencies of two input signals. In one method the two input signals are simply connected in series or in parallel and applied between two input terminals such as the base and emitter of a transistor. If the input-output relationship for the transistor is linear, these two signals are amplified independently and there is no output at any frequency other than frequencies of the two input signals. To produce interaction between the two original signals, thus obtaining an output at the difference frequency, it is essential that the transistor should have a non-linear characteristic, i.e. should behave as a detector.

Stages of this type are known as *additive mixers* and they are usually biased near the point of collector-current cut off to achieve the non-linearity essential for their action.

In another type of mixer the received signal and local-oscillator output are in effect multiplied. This is achieved by applying the two signals to two control electrodes of a transistor which is so designed that the signal on one electrode controls the gain of the signal applied to the other electrode. In a mixer of this type there is no need for non-linearity, the output at the difference frequency being produced directly from multiplication of the two signals. Such mixers are termed *multiplicative*.

Field-effect transistors can be used as additive or multiplicative mixers. The $I_d - V_g$ characteristic of an f.e.t. has square-law curvature and thus additive mixing can be achieved by applying both signals to the gate. Field-effect transistors can have two input terminals: these may both be gates or one may be a gate and the other a base connection. Such transistors may be used as multiplicative mixers by connecting the input to one terminal and the oscillator output to the other.

Bipolar transistors have only a single control electrode and are therefore used as additive mixers, and one possible circuit is given in Fig. 12.5. This also gives the circuit of the oscillator stage; the combination of the two constitutes a frequency-changer stage.

In this circuit the two input signals are connected in series and applied to the base-emitter circuit of the mixer. The operating point

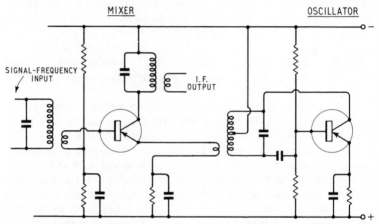

Fig. 12.5. One possible circuit for a two-transistor frequency changer

is swept over the characteristic by the local oscillation, and the efficiency of the mixing process depends on the amplitude of oscillation injected into the emitter circuit. For small oscillation amplitudes the conversion gain, i.e. the ratio of i.f. signal output to r.f. signal input, increases linearly with oscillation amplitude but levels off at a particular amplitude and then falls. Increase in oscillator amplitude beyond this point causes little change in gain.

In a receiver in which the oscillator has to operate over a range of frequency, some variation in oscillator amplitude is inevitable. In order to prevent such variations causing large changes in conversion gain the oscillator amplitude is usually designed so that it is greater than the minimum critical value which gives maximum gain.

Self-Oscillating Mixers

It is quite possible to arrange for a single transistor to perform the functions of mixing and of oscillation, and the conversion gain of such a circuit can be almost equal to that of circuits in which separate transistors are used for the two purposes.

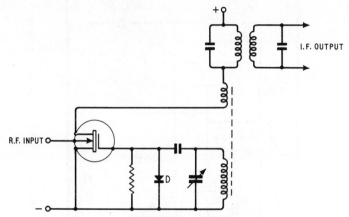

Fig. 12.6. Self-oscillating mixer circuit using an i.g.f.e.t.

One circuit for a self-oscillating mixer using an i.g.f.e.t. is illustrated in Fig. 12.6. The gate and drain are connected in a Reinartz oscillator circuit and the diode D provides gate bias by rectifying the generated r.f. oscillation. The r.f. input is connected to the base of the transistor

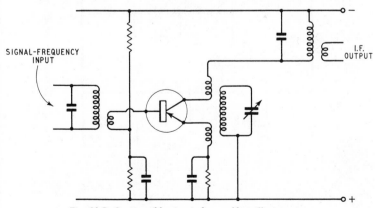

Fig. 12.7. One possible circuit for a self-oscillating mixer

210

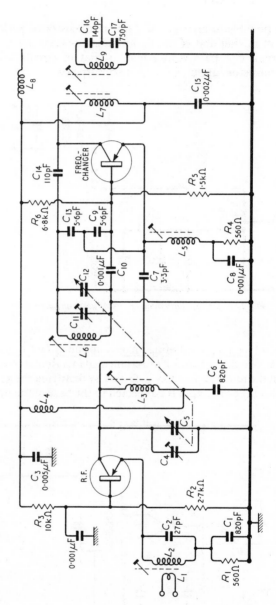

Fig. 12.8. Circuit diagram of a transistorised v.h.f. tuner including r.f. amplifier and self-oscillating mixer

which is not so sensitive as the gate but nevertheless enables mixing to be achieved.

Another circuit for a self-oscillating mixer using a bipolar transistor is given in Fig. 12.7. This may be regarded as a Reinartz oscillator of the type shown in Fig. 11.4 in which the base circuit is tuned to accept the signal-frequency input and the output circuit is tuned to select the difference-frequency output. The oscillator signal and the signal-frequency input are connected in series between base and emitter to enable the transistor to operate as an additive mixer. Circuits of this type are commonly employed in transistor superheterodyne receivers for a.m. reception.

V.H.F. TUNER

Fig. 12.8 shows a form of self-oscillating mixer suitable for use in v.h.f. receivers. The transistor operating conditions are set at a value suitable for a self-oscillating mixer by the potential divider R_5R_6 which determines the base potential and by the resistor R_4 which determines the emitter current. The transistor circuit is based on that of a Colpitts' oscillator, the two fundamental capacitors C_9 and C_{13} being connected in series between collector and base with the centre point connected to emitter. C_{12} is the oscillator tuning capacitor and C_{11} a trimmer. C_{12} is ganged with C_5 which tunes the signal-frequency inductor L_3 in the collector circuit of the r.f. amplifier. It is desirable to earth the moving vanes of C_5 and C_{12} but if this is done the base of the oscillator is at zero r.f. potential. The emitter cannot also be at zero r.f. potential (as in many Colpitts' oscillators) because C_9 would be short-circuited and oscillation would be impossible. Some reactance is therefore provided in the emitter circuit by the inclusion of inductor L_5. The signal-frequency input to the mixer can now be injected into the emitter circuit via C_7 to enable mixing to take place in the base-emitter diode of the frequency changer. The inclusion of L_5 in the emitter circuit of the frequency changer provides negative feedback which reduces the gain of the stage at the intermediate frequency. The feedback is therefore reduced to a minimum by arranging that L_5 resonates with C_8 at the intermediate frequency. L_5 is, therefore, made variable to permit this adjustment.

The i.f. output of the frequency changer is selected by the transformer L_7L_9, both primary and secondary windings of which are tuned to the intermediate frequency, usually 10·7 MHz. The primary winding is tuned by the 110-pF capacitor C_{14} which is connected

212

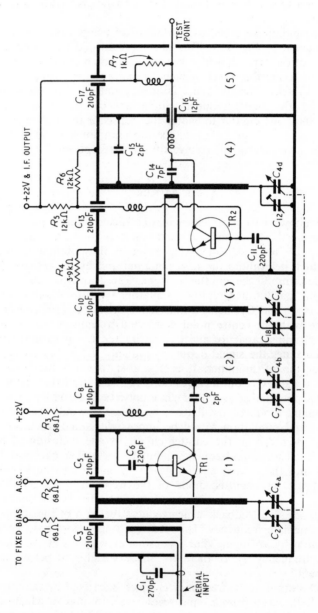

Fig. 12.9. Circuit diagram of a u.h.f. tuner

in series with L_6 and C_{15} across L_7. Both L_6 and C_{15} have negligible reactance at 10·7 MHz and thus C_{14} is effectively in parallel with L_7.

Chapter 10 shows how the internal base-collector capacitance of a transistor limits the gain available from a common-emitter stage. Feedback via this capacitance becomes more serious as frequency rises: for this reason the stable gain obtainable at 10·7 MHz is considerably less than at 465 kHz. At v.h.f. and u.h.f. the gain of a common-emitter stage is so reduced by internal feedback that a common-base amplifier is usually preferred.

Fig. 12.8 includes such a stage. This is stabilised by the potential divider $R_2 R_3$ and the emitter resistor R_1. The low input resistance of the common-base stage is no disadvantage in a v.h.f. receiver because it can be used to damp the tuned circuit $L_2 C_2$ to give coverage of the entire band (87·5 to 100 MHz), so avoiding the necessity for variable tuning in this part of the receiver. The tuning is preset to the centre of the band say at 94 MHz.

U.H.F. TUNERS

Fig. 12.9 gives the circuit diagram of an r.f. stage and frequency changer for operation on Bands IV and V (470 MHz to 854 MHz): such a circuit is used in the tuner of a 625-line television receiver.

At such frequencies it is not possible to use conventional wound inductors in tuned circuits and lengths of transmission line are used instead. A $\lambda/4$ length of transmission line, if short-circuited at one end, presents a very high resistance at the open end: this is a resonant condition and the open-end impedance is similar to that of a parallel-tuned circuit at resonance. If the line is shortened to less than $\lambda/4$ or $3\lambda/4$ it becomes inductive and a variable capacitor can be connected across the open end to tune it. This is the technique used in u.h.f. tuners: a box chassis is divided by partitions into rectangular sections each of which is used as the outer conductor of a trough line, a stout wire forming the centre conductor. Tappings can be made on the centre conductor to give a suitable resistance for matching purposes or a short length of conductor can be placed parallel to it to form a transformer.

Two such conductors are used in the first trough line in Fig. 12.9: one provides a suitable terminating resistance for the coaxial cable input; the other provides matching to the emitter circuit of the common-base amplifier TR1. The second trough line is fed via choke-capacitance coupling from TR1 collector. This line forms a bandpass filter with the third trough line, coupling being provided by an aperture in the partition between the lines. TR2 is

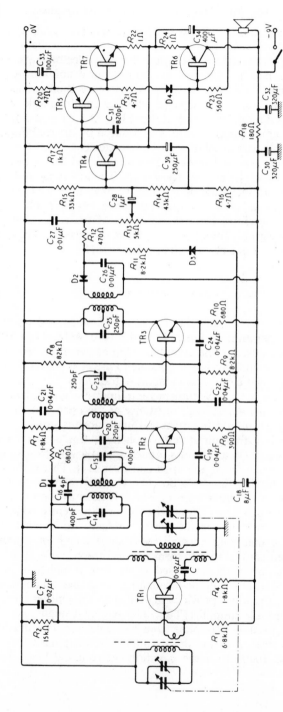

Fig. 12.10. Simplified circuit diagram of an a.m. sound receiver

also a common-base stage, choke-capacitance coupled to the fourth trough line and the positive feedback required for oscillation is provided by collector-emitter capacitance. TR2 emitter receives a signal-frequency input via the coupling trough line (3) and an oscillator input via the coupling to trough-line (4): TR2 is an additive mixer as well as oscillator. The inductor in the fifth compartment matches TR2 output to the input of the i.f. amplifier. The four trough lines are tuned by the sections of the four-gang capacitor C_4.

A.M. SOUND RECEIVER

Fig. 12.10 gives the circuit diagram for a battery-operated a.m. sound receiver designed for silicon planar transistors in the f.c. and i.f. stages. The circuit is intended for m.w. and l.w. reception but for simplicity all wave-change switching has been omitted.

The a.f. amplifier, detector, i.f. amplifier and f.c. stages are all similar to those previously described but the measures adopted to maintain d.c. stability of the silicon transistors in spite of falling battery voltage merit some description. Because of the high offset voltage of these transistors, if the conventional circuit using a base potential divider and emitter resistor is employed, a large fraction of the supply voltage must be sacrificed: moreover large-value emitter resistors are required and these degrade a.g.c. performance. A.g.c. is not used on the f.c. stage and here the conventional circuit is used, 2 V being lost across the 1·8 kΩ emitter resistor R_4.

Both i.f. stages are stabilised by the same potential-divider circuit and the stability is improved by the inclusion of a forward-biased diode D3 as described in Chapter 6. To provide a.g.c. the potential divider is returned to the detector circuit and thus the complete chain includes R_8, R_9, D3, R_{11} and D2. TR2 is fed from R_9, D3 junction and has a low-value emitter resistor R_6 (390 Ω) so that automatic gain control has a wide range: TR3 is fed from R_8, R_9 junction and has a larger value of emitter resistor R_{10} (680 Ω). This technique is adopted to limit the extent of control on TR3 so that this transistor, even on very strong signals, has sufficient collector current to provide the detector with an adequate input signal.

A further contribution to a.g.c. is provided by the circuit R_5 D1. For weak signals TR2 takes, say 1 mA collector current and there is a voltage of 2 across R_7. Thus D1 is reverse-biased and does not conduct until the signal across the primary of the 1st i.f. transformer approaches 2 V peak and this, of course, is most unlikely. On very strong signals, however, TR2 is reverse-biased so that the collector current and the voltage across R_7 fall to a very low value. Even a small signal across the transformer primary now causes D1 to conduct: this applies considerable damping to the transformer and reduces the signal level. R_5 D1 thus give protection against overloading on very strong signals.

A.M.–F.M. SOUND RECEIVERS

F.M. reception is in general far superior to a.m. It permits reproduction of frequencies up to 15 kHz without introducing problems of selectivity and, provided adequate signal strength is available, can eliminate interference to provide a quiet background. Thus transistor receivers, even portables, often cover the f.m. as well as the a.m. broadcast bands. A common technique is to include the 10·7-MHz f.m. i.f. transformer windings in series with those of the 465-kHz a.m. i.f. transformers. The gain at 10·7 MHz is however much lower than at 465 kHz and the a.m. frequency changer is converted by the waveband switch to a third 10·7-MHz amplifier to provide adequate i.f. gain. A separate 2-transistor v.h.f. tuner such as that illustrated in Fig. 12.8 is brought into use for f.m. reception. The waveband switching must also select the appropriate form of detector.

TELEVISION RECEIVERS

Television receivers, particularly colour receivers, are complex aggregates of electronic circuits and it is possible to devote entire books to the use of transistors in such receivers. Information on many of the transistor circuits used in television receivers can be found in this book under the headings of tuner (r.f. and f.c. stages), i.f. amplifiers, detectors (a.m. and f.m.), video amplifiers, colour-difference amplifiers, a.f. amplifiers, line and sawtooth generators.

CHAPTER 13

Pulse Generators

INTRODUCTION: THE TRANSISTOR AS A SWITCH

So far in this book we have discussed the use of transistors as amplifiers of pulses or sinusoidal signals. In class-A and class-B amplifiers the transistor is used as a linear amplifier in which every change in input signal brings about a corresponding change in output signal. In this type of operation (typical of analogue equipment) the shape of the input–output characteristic is all-important and any deviations from linearity must be compensated by negative feedback or other means to give an acceptable performance.

In pulse or digital circuits, however, transistors are used in a completely different type of operation in which the shape of the input–output characteristic is of little significance. The transistor is, in fact, used as a switch with only two states usually termed *on* and *off*. In the *on* state the transistor is conductive, having a significant collector (or drain) current, the collector-emitter (or drain-source) voltage being very low. In the *off* state the transistor is non-conductive, having negligible collector (or drain) current, the collector-emitter (or drain-source) voltage being practically equal to that of the supply. In both states dissipation in the transistor is low, in the *on* state because of the low collector (or drain) voltage and in the *off* state because of the low collector (or drain) current. In the *on* state, however, there is appreciable dissipation in the base circuit of a bipolar transistor for this is usually fed via a series resistor. There is no corresponding dissipation in the f.e.t. which gives it one advantage over the bipolar transistor in digital applications.

Stable and Unstable States

Sometimes in a digital circuit a transistor will remain in one of its two possible states indefinitely unless it is compelled to change state by an externally-applied signal: such a state is termed *stable*. Alternatively a circuit can be such that a transistor will remain in a given state only for a limited period after which without any help from external sources it automatically reverts to the other state: such states are termed *unstable*.

So far we have spoken of the two possible states of a single transistor but a circuit containing several transistors may also possess stable and/or unstable states. In both states some of the transistors may be on whilst others are off but it is still true that the states of the circuit as a whole may be stable or unstable.

Bistable, Monostable and Astable Circuits

In some pulse circuits both states are stable. Such a circuit is termed *bistable* and it requires an external triggering signal to compel it to leave one state and enter the other. A further external signal is then required to compel the circuit to return to the original state again. Two triggering signals are thus necessary for each cycle of operation of the circuit which therefore has applications as a counter or frequency divider.

In other circuits one state is stable and the other unstable. Such a circuit is termed *monostable* and it always reverts to its stable state. An external signal is needed to make the circuit enter the unstable state but, after such stimulus, the circuit automatically reverts to the stable state. The time taken to reach the stable state after triggering can be given any desired value within wide limits by appropriate choice of time constants. Circuits of this type are often used to give accurately-timed delays and are sometimes called delay generators.

In a further class of circuits both states are unstable. Such a circuit is termed *free-running* or *astable* and will not remain permanently in either state. Without need of external signals an astable circuit automatically switches between the two states at a frequency determined by the time constants of its circuitry. Although an astable circuit does not need external signals to make it operate, it can readily be synchronised at the frequency of a recurrent external signal.

Some circuits can be made bistable, monostable or astable by simple changes in circuitry, sometimes even by altering bias conditions. One circuit of this type is the multivibrator.

MULTIVIBRATOR: BISTABLE

The basic circuit for a collector-coupled bi stable multivibrator is given in Fig. 13.1. This shows two transistors with resistive collector loads, each collector being connected to the other base by a resistive potential divider. The arrangement is such that when

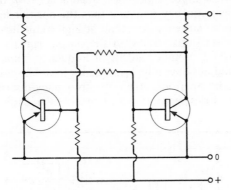

Fig. 13.1. Basic circuit of bistable multivibrator using bipolar transistors

one transistor is conducting, its collector potential ensures that the other transistor is cut off. The two possible states of the circuit are thus:

(1) TR1 on and TR2 off.
(2) TR2 on and TR1 off.

If the circuit is initially in state (1), an external signal will cause it to enter state (2) and the change of state is accomplished with great rapidity. The speed of transition is due to the positive feedback inherent in the circuit: this feedback is obvious if the circuit is regarded as that of a two-stage direct-coupled amplifier with the output voltage returned to the input. A second external signal applied to the circuit will cause it to return to state (1) again.

During these changes of state the collector potentials of TR1 and TR2 alternate between a negative value when the transistor is cut off and almost zero when it is fully conducting.

If a recurrent triggering signal is applied to the circuit, the voltage waveform generated at each collector is approximately rectangular in form. Moreover, the frequency of the rectangular signal is half that of the triggering signal, enabling such circuits to be used in frequency dividers or binary counters.

Speed-up Capacitors

In the simple form illustrated in Fig. 13.1 the waveforms generated at the collectors during changes of state have poor rise times. This is because the circuit coupling each collector to the other base consists of a series resistor feeding into the parallel RC combination of the transistor input impedance. Such a network has a response which falls as frequency rises and thus the positive feedback in the circuit is more marked at low than at high frequencies. To achieve steep edges (small rise times) in the output signals the degree of feedback should be independent of frequency: thus the coupling circuits should not have a high-frequency loss. The coupling circuits can be made aperiodic (i.e. non-frequency-discriminating) by shunting the series resistors with capacitors as shown in Fig. 13.2

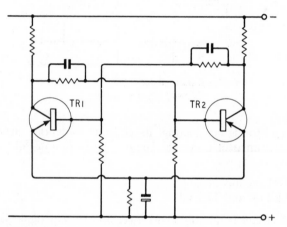

Fig. 13.2. Transistor bistable multivibrator with speed-up capacitors

and the condition for an aperiodic response is that the time constant of the coupling resistor-coupling capacitor combination should equal that of the transistor input impedance. If a smaller value of capacitor is used the rise time of the multivibrator output signals is not improved as much as is possible: if a larger value of capacitor is used the capacitance placed in parallel with the collector load becomes significant and it slows up the speed with which the collector potential approaches the negative supply potential when the collector current is cut off: in other words the rise time of the negatively-going collector potential changes is impaired.

To avoid the necessity for the separate positive supply for the bases shown in Fig. 13.1, the two emitters may be bonded and connected to the earth line as shown in Fig. 13.2. In this circuit the emitter current of the conductive transistor ensures that the other is cut off by making its emitter potential more negative than its base potential. It is advisable to decouple the common-emitter resistance to preserve steep edges in the output waveform.

Triggering signals applied to a bistable transistor circuit to interchange the states of conduction or non-conduction can be applied to the base or the emitter circuit. It is preferable, however, to apply signals to the base because this takes advantage of the β of the transistors to amplify the triggering signals. If pnp transistors are used, a positive-going signal is required to cut off a conductive transistor or a negative-going signal to make a cut-off transistor conductive. Of these two alternatives the first is in general preferable because it requires a smaller voltage signal, there being no large bias to overcome. Moreover, a triggering signal applied to a conductive transistor is amplified by the transistor causing a much larger triggering signal to be applied to the second transistor.

In counter circuits it is usual to inject a series of unidirectional

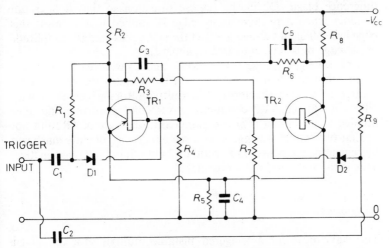

Fig. 13.3. The circuit of Fig. 13.2 with diode triggering gate

pulses into a bistable circuit and some means is required of directing the pulses alternately to the base circuits of the two transistors. One method of achieving this is by the use of the diode gate circuit illustrated in Fig. 13.3. The input signals are applied to the two capacitors C_1 and C_2 and then to the two base circuits via diodes

D1 and D2. The switching action of the diodes can be followed from the following section.

Diode Gate Circuit

Suppose TR1 is conductive. The collector current is a maximum and the collector potential is very low so that there is little difference between the collector and base potentials. The diode D1 is connected between collector and base (via R_1) and there is little potential difference across the diode. The diode is so connected that any positive-going edge applied to capacitor C_1 makes the diode conduct, i.e. low-resistance and the edge is conducted to the base of TR1. The edge cannot, however, reach the base of TR2: this transistor is cut off and its collector current is a minimum (consisting of leakage current I_{CBO} only). The collector potential of TR2 is very high and, except for the voltage drop across R_8 due to leakage current, would be equal to the supply voltage. There is thus a maximum potential difference across the diode D2 which is connected between collector and base (via R_9). Moreover this potential difference biases D2 in the reverse direction so that it does not conduct when a positive-going edge is applied to C_2 unless this edge has sufficient voltage to offset the static bias (and the amplitude of the edge must be controlled to avoid this).

A positive-going edge applied to C_1 and C_2 thus reaches the base of TR1 but not that of TR2. This edge initiates the regenerative change of state characteristic of multivibrators which ends with TR1 cut off and TR2 conducting. A second positive-going edge applied to the circuit is now conducted to the base of TR2 but not to that of TR1. This triggers off a second change of state, hastened by regeneration, which ends with TR2 cut off and TR1 conductive. This is the state originally postulated.

Waveforms

The waveforms for a triggered bistable multivibrator are given in Fig. 13.4. At t_1 TR1 is conductive and its base and collector potentials are slightly negative with respect to emitter potential. At the same instant TR2 is cut off: its base potential is considerably positive with respect to emitter potential and the collector potential is almost at supply potential $-V_{cc}$. When $t = t_2$ a triggering pulse is received by TR1 and interchanges the states. At $t = t_3$ a second triggering edge is received (by TR2 this time) and interchanges the

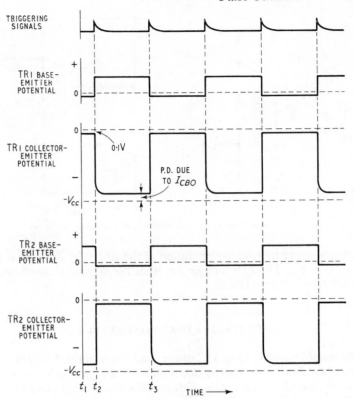

Fig. 13.4. Base and collector waveforms for the bistable multivibrator of Fig. 13.3

states again. Two triggering signals are thus required to produce one complete rectangular wave from TR1 or TR2 collector circuit.

Complementary Bistable Multivibrator

The transistors in a multivibrator must provide the two essentials of phase inversion and gain. A bistable circuit can therefore consist of a pnp and an npn transistor. Such a complementary multivibrator has the unusual feature that in one state both transistors are on and in the other both are off. Direct coupling can be used between collectors and bases and as a result the circuit is particularly simple containing only five resistors as shown in Fig. 13.5(a): here the output is taken from the emitter circuit of the npn transistor. It is

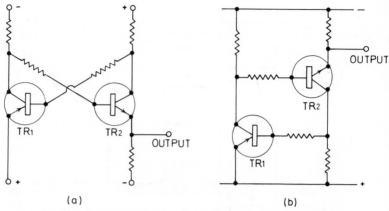

Fig. 13.5. A complementary bistable multivibrator shown in two forms

perhaps easier to visualise operation of the circuit by drawing it as shown in Fig. 13.5(b) although this loses the familiar crossed lines of the multivibrator circuit.

MULTIVIBRATOR: MONOSTABLE

The bistable circuit of Fig. 13.2 can be made monostable by replacing one of the direct inter-transistor couplings by a capacitance coupling and by suitable choice of bias. One possible circuit for a monostable collector-coupled multivibrator is that shown in Fig. 13.6. TR1 is

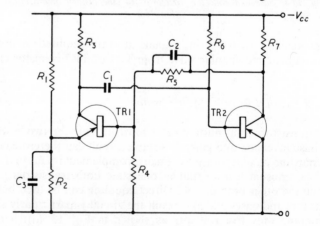

Fig. 13.6. Transistor monostable multivibrator

biased by the potential divider $R_1 R_2$ but TR2 has no bias components other than R_6 which connects the base to the negative supply line. Clearly any positive potential applied to TR2 base (to cut this transistor off) cannot remain there indefinitely but will leak away as C_1 discharges and thus the circuit will always revert to the state in which TR2 is on (and TR1 therefore cut off). Thus we have

stable state: TR1 off and TR2 on

from which it follows that

unstable state: TR2 off and TR1 on.

The operation of the circuit is illustrated by the waveforms of Fig. 13.7. At time t_1 the circuit is assumed here to be in the stable state.

At t_2 TR1 is made conductive by applying a negative-going edge to its base or preferably (as shown in Fig. 13.7) by applying a positive-going edge to TR2 base. The collector potential of TR1 makes a positive excursion (AB in Fig 13.7). The voltage across a capacitor cannot be changed instantaneously and this edge is transferred without loss in amplitude by C_1 to the base of TR2 (ab), cutting TR2 off. The circuit is now in the second of its two possible states (TR1 conductive and TR2 cut off) but it cannot remain in it permanently because C_1 begins immediately to discharge through R_6 and the output circuit of TR1. As the discharge proceeds the potential at the base of TR2 becomes less positive (bc) until a point (c) is reached at which TR2 begins to conduct. This causes the collector potential of TR2 to move positively, generating an edge CD which is transferred simultaneously, without change in amplitude, to TR1 base (cd) cutting TR1 off. This in turn causes the collector potential of TR1 to move negatively which accelerates the onset of conduction in TR2.

The process becomes regenerative, in fact, as soon as the overall gain of the two transistor stages exceeds unity, and the change of state is then very rapid. After the change, TR1 is off and TR2 on. The potential at TR1 base now falls to near zero as C_2 discharges through R_4, R_5 and the output circuit of TR2. The potential at TR1 collector moves exponentially towards $-V_{cc}$ as C_1 charges through R_3. When these two changes have ceased the circuit rests and no further action occurs unless further triggering signals are received: in other words this is the stable stage again.

When a monostable multivibrator is used as a delay generator, the delay is equal to the period of the unstable state and we thus need to be able to calculate the component values needed to produce a wanted delay. This can be achieved as follows.

Duration of Unstable State

The duration of the unstable state is equal to the time taken for the potential at the base of the non-conductive transistor to fall to just beyond cut-off value.

First consider the conditions in the circuit in the stable state: TR1 is cut off and TR2 conductive. If TR1 is a silicon transistor its collector potential would be equal to that of the supply $-V_{cc}$. TR2 is conductive and its base potential is at the cut-off voltage $-V_{be}$. Thus the voltage across C_1 is equal to $(V_{cc} - V_{be})$, the right-hand plate being positive with respect to the other.

The unstable period begins when TR1 is suddenly switched on. The collector current rises at this instant to a value which is sufficient with proper design to bring the collector potential very nearly to TR1 emitter potential V_{e1}. The change in TR1 collector potential is thus $(V_{cc} - V_{e1})$ and this is transferred by C_1 to TR2 base, causing the potential here to rise from $-V_{be}$ to $(V_{cc} - V_{e1} - V_{be})$. C_1 now begins to discharge through R_6 and the output circuit of TR1 but the resistance of this circuit (with TR1 conductive) is small compared with R_6 and can be neglected. If the discharge were completed TR2 base potential would fall from its initial (positive) value of $(V_{cc} - V_{e1} - V_{be})$ to zero and would then reverse in polarity and reach the (negative) supply voltage of $-V_{cc}$. However, as soon as TR2 base potential reaches $-V_{be}$, the cut-off value, TR2 begins to conduct and the unstable period is abruptly terminated.

The rate of discharge of C_1 is determined by the discharge current. This flows in R_6 and thus the current can be measured by the voltage across R_6. At the beginning of the discharge TR2 base has a potential of $(V_{cc} - V_{e1} - V_{be})$ and as the supply voltage is $-V_{cc}$, the voltage across R_6 is $(2V_{cc} - V_{e1} - V_{be})$. At the moment when TR2 becomes conductive the voltage across R_6 is $(V_{cc} - V_{be})$. The duration of the unstable state is approximately equal to the time taken for this fall of voltage to occur in a circuit of time constant R_6C_1. In general the fall of voltage is given by

$$V_t = V_o e^{-t/R_6 C_1}$$

where V_t is the voltage after a time t and V_o is the initial voltage. This may be written in the form

$$\log_e \frac{V_o}{V_t} = \frac{t}{R_6 C_1}$$

from which

$$t = R_6 C_1 \log_e V_o/V_t$$

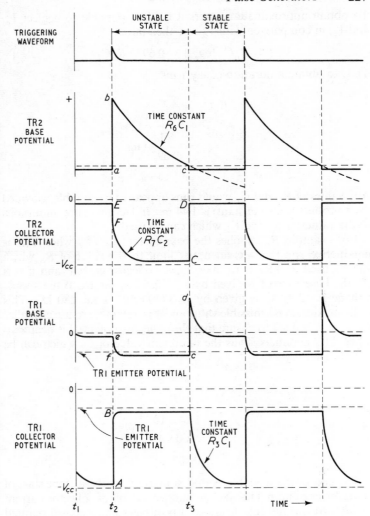

Fig. 13.7. Base and collector potential waveforms for the multivibrator of Fig. 13.6

Substituting for V_o and V_t

$$t = R_6 C_1 \log_e (2V_{cc} - V_{e1} - V_{be})/(V_{cc} - V_{be})$$

As typical practical values in a circuit using silicon transistors we may have $V_{cc} = 12$ V, $V_{ce} = 3$ V and $V_{be} = 0.7$ V. For these

$$t = R_6 C_1 \log_e 1.8 = 0.587 R_6 C_1$$

To obtain approximate answers it is often possible to neglect V_{e1} and V_{be} in comparison with V_{cc}: we then have

$$t = R_6 C_1 \log_e 2 = 0.6931 \, R_6 C_1$$

Thus to obtain a duration of say 1 ms

$$R_6 C_1 = \frac{t}{0.6931}$$

$$= \frac{1}{0.6931} \text{ ms}$$

$$= 1.44 \text{ ms}$$

Any values of R_6 and C_1 would thus appear to be suitable provided their product (time constant) is 1.44 ms. In fact there are limitations to the values of R_6 and C_1 which can be used.

For example, R_6 supplies the base current to TR2 when TR2 is conductive and this sets an upper limit to the value of R_6 which can be used. If TR2 is to take 2 mA collector current and if β is 50, the base current is given by 2/50 mA, i.e. 40 μA. If the supply voltage is 12 V R_6 is given by $11.3/(40 \times 10^{-6})$, i.e. 280 kΩ. This is the maximum permissible value and it is advisable to use a smaller value, say, 200 kΩ to ensure that the collector potential is near zero when TR2 conducts. Thus the minimum value of C_1 which can be used is given by

$$C_1 = \frac{1.44 \times 10^{-3}}{R_6}$$

$$= \frac{1.44 \times 10^{-3}}{200 \times 10^3} \text{ F}$$

$$= 0.007 \text{ μF}$$

The end of the unstable period is marked by the abrupt start of collector current in TR2 and the sudden cut-off of collector current in TR1. At this instant, therefore, a positive-going step is generated at TR2 collector and a negative-going step at TR1 collector. Either of these steps could be used as the output of the circuit but the output at TR2 collector is, in general, preferred because it has a shorter rise time. This is borne out by the following calculations.

When TR2 is turned on, its collector potential goes positive and cuts TR1 off by driving its base positive. There is little change in voltage across the coupling capacitor C_2 and the only capacitance which has to be charged or discharged by the collector current of TR2 is the output capacitance of TR2 with possibly a contribution of capacitance from TR1 input. The relationship between the voltage across a capacitor and the current flowing into it is given by

$$V = \frac{Q}{C} = \frac{1}{C} \int i \, . \, dt$$

from which

$$\frac{dV}{dt} = \frac{i}{C}$$

Suppose the capacitance at TR2 collector is 20 pF and that the collector current of TR2 is 2 mA. We have

$$\frac{dV}{dt} = \frac{i}{C}$$

$$= \frac{2 \times 10^{-3}}{20 \times 10^{-12}}$$

$$= 10^8 \text{ V/s}$$

$$= 100 \text{ V/}\mu\text{s}$$

If the supply to TR2 is 12 V the collector potential excursion is limited to this value and occupies only approximately 0·1 µs. The rise time of this potential step is thus of the order of 0·1 µs. This is perhaps an optimistic value because we have neglected any effect due to TR2 collector resistor R_4 which tends to slow up the discharge of TR2 collector capacitance by charging it from the supply.

Alternatively, of course, the potential step generated at TR1 collector may be used as output. When TR1 is cut off at the end of the unstable period, the collector potential approaches that of the negative terminal of the supply. This tends to drive the base of TR2 negative but as soon as TR2 starts to conduct, its input resistance becomes very low and its base potential is thus effectively stabilised near emitter potential. Thus TR1 collector potential can only go negative as C_1 charges from the supply through R_3. This change in potential is governed by the time constant R_3C_1 and the rise time is given approximately by $2R_3C_1$. A suitable value for R_3 is 5 kΩ because this permits a collector current of 2 mA with a supply of 12 V (2 V being assumed lost across TR2 emitter resistor). C_1 has already been calculated as 0·007 µF and thus we have

$$\text{rise time} = 2R_3C_1$$

$$= 2 \times 5 \times 10^3 \times 0 \cdot 007 \times 10^{-6} \text{ s}$$

$$= 70 \text{ µs}$$

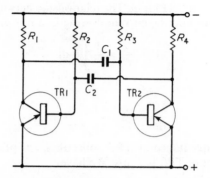

Fig. 13.8. Transistor astable multivibrator

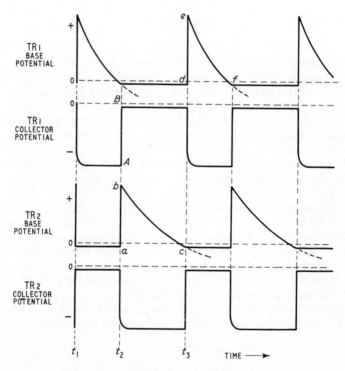

Fig. 13.9. Base and collector waveforms for the astable multivibrator of Fig. 13.8.

which is considerably greater than the rise time of the step at TR2 collector.

Hole Storage

The difference between the rise times of the voltage steps at TR1 and TR2 collectors is in practice likely to be greater than suggested by the above calculations. This is because of an effect, known as hole storage, which tends to degrade the rise time of the collector-potential step when a transistor is cut off. Hole storage is an effect, internal to the transistor, which causes the collector current to persist for a few microseconds after the emitter current has been cut off. This occurs in transistors which have been driven hard into conduction so that the collector-emitter potential is nearly zero. In these circumstances the emitter injects more charge carriers (holes in a pnp transistor) into the base region than are required to give the collector current, and the excess carriers are stored ready to be swept into the collector region to prolong the collector current for a short period after the emitter current has been cut off. This hole-storage effect can be avoided by so designing the circuits that the collector-emitter potential does not approach zero but is limited to a value such as 1 V: this can be achieved by the use of diodes.

MULTIVIBRATOR: ASTABLE

The bistable circuit of Fig. 13.2 can be made astable by replacing both of the direct inter-transistor couplings by capacitance couplings and by suitable choice of bias. One possible circuit for an astable collector-coupled multivibrator is given in Fig. 13.8.

No emitter bias is used and both bases are returned via resistors to the negative supply line. The behaviour of the circuit will be illustrated by the waveforms shown in Fig. 13.9. At t_1 TR1 is cut off by a positive signal at the base and the collector potential is a negative maximum. At the same time TR2 is conducting, having minimum negative values of base and collector potentials. TR2 is in a stable state but TR1 is not, because the potential at its base is moving negatively as C_2 discharges through R_2. At t_2 the potential at TR1 base is low enough for TR1 to start conducting. As explained for the monostable multivibrator, this starts a regenerative action which causes TR1 to become abruptly conductive. The resultant steep positive-going voltage (AB) at TR1 collector is transferred by C_1 to the base of TR2 (ab), cutting TR2 off. There now follows the exponential change in base potential (bc) previously described. It keeps TR2 cut off for a period given approximately by $0.69R_3C_1$ which terminates at time t_3 when TR2 becomes abruptly conductive. TR2 is now in a stable state but TR1 is not, being cut off by a positive edge (*de*) which decays exponentially (*ef*). This keeps TR1 cut off for a period given approximately by $0.69R_2C_2$. This is the circuit condition assumed initially. Thus the cycle continues, the period of oscillation T being equal to the sum of the durations of the two unstable periods

$$T = 0.69(R_3C_1 + R_2C_2)$$

If $R_2 = R_3 = R$ and $C_1 = C_2 = C$ the multivibrator is symmetrical and generates square waves (equal mark space ratio) at both collectors. The period of oscillation is given by

$$T = 1.38RC$$

The free-running frequency of such a multivibrator is given by

$$f = \frac{1}{T}$$

$$= \frac{1}{1.38RC}$$

A multivibrator required to be synchronised at the line frequency (approximately 15 kHz) of a 625-line television system would be designed to have a natural frequency somewhat lower than this— say 10 kHz. We have already seen that suitable values for the load resistors and base resistors are 5 kΩ and 200 kΩ: we have now to determine a suitable value for the coupling capacitors. From the above expression we have

$$C = \frac{1}{1.38Rf}$$

$$= \frac{1}{1.38 \times 2 \times 10^5 \times 10^4} \, \text{F}$$

$$= 360 \, \text{pF approximately}$$

SYNCHRONISING OF MULTIVIBRATORS

Bistable and monostable multivibrators are triggered by signals which cause the circuit to leave a stable state. Such a technique cannot be applied to astable circuits because these have no stable states. The synchronising signals injected into an astable circuit are designed to terminate the unstable periods earlier than would occur naturally. It follows that the natural frequency of the circuit must be lower than the frequency of the synchronising signals.

To terminate the unstable periods unnaturally early the synchronising signals can take the form of negative-going pulses applied to one or both of the bases of pnp transistors. The amplitude of the edges is important. If the natural frequency of the multivibrator is small compared with the frequency of the synchronising signals, a small-amplitude signal may cause the multivibrator to run at a simple fraction, say $\frac{1}{5}$th of the sync frequency. Increase in the amplitude of the synchronising signal may cause the multivibrator frequency to jump to $\frac{1}{4}$th of the synchronising frequency. Further increase may cause the frequency to jump to $\frac{1}{3}$rd the synchronising frequency, etc. Thus a synchronised multivibrator can be used as a frequency divider but close control of the synchronising signal amplitude is necessary to obtain a consistent division ratio.

EMITTER-COUPLED MULTIVIBRATOR

All the multivibrators described above have employed two collector-to-base coupling circuits. One of these can be replaced by an emitter-to-emitter coupling without significantly altering the principles of operation and Fig. 13.10 gives the circuit diagram of a bistable emitter-coupled multivibrator. When TR2 is turned on, its emitter current, in flowing through the common-emitter resistor R_e, makes the emitter of TR1 negative with respect to the fixed base potential and thus cuts TR1 off. The effect of R_e is hence similar to that of a direct coupling between TR2 collector and TR1 base.

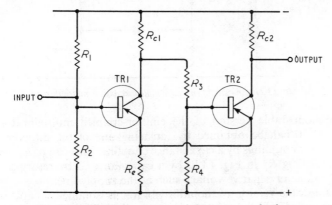

Fig. 13.10. A bistable emitter-coupled transistor multivibrator

One advantage of this circuit over the collector-coupled version is that the collector of TR2 plays no part in the mechanism of operation. It is free and can provide an output, the loading on which has no significant effect on the performance of the circuit.

Collector-coupled multivibrators can be symmetrical but emitter-coupled types cannot. This can easily be seen for in one of the two possible states of the circuit the voltage across R_e must be less than that across R_2 to give the negative base-emitter voltage which keeps TR1 conductive. In the other state, however, the voltage across R_e must be greater than that across R_2 to give the positive base-emitter voltage which keeps TR1 cut off. In the first state the only current in R_e is that due to TR1 whereas in the second state the current in R_e is due entirely to TR2. It follows that TR2 must take a larger collector current (when conductive) than TR1 (when conductive) and the circuit is hence asymmetrical.

The circuit can be triggered by positive-going or negative-going signals applied to TR1 or TR2 base and is sometimes used as a limiter by applying, say, a sinusoidal input to TR1 base. The output is then rectangular at the frequency of the input with an amplitude substantially independent of that of the input provided this exceeds a certain minimum value.

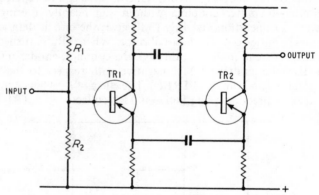

Fig. 13.11. An astable emitter-coupled transistor multivibrator

A monostable version of the emitter-coupled multivibrator of Fig. 13.10 can be obtained by replacing one of the direct inter-transistor couplings by a capacitance coupling. This can be achieved by replacing R_3 in Fig. 13.10 by a capacitor and by returning R_4 to a source of negative voltage such as the supply line.

An astable version of the multivibrator is obtained if capacitors are included in both inter-transistor couplings. This may be achieved by modifying the collector-to-base coupling as for the monostable circuit and in addition using individual emitter resistors coupled by a capacitor as shown in Fig. 13.11.

BLOCKING OSCILLATOR

As we have seen, a multivibrator consists essentially of two transistors so connected that the output of each is coupled to the input of the other. Each transistor introduces a phase inversion, a condition essential to give the positive feedback which accelerates the changes of state. If one of the transistors is replaced by a transformer connected so as to give phase inversion, the resulting circuit will still generate pulses. The circuit derived in this way is termed a blocking oscillator and is extensively employed as a pulse generator and as a sawtooth generator. Two forms of blocking oscillators are in common use, an astable or free-running circuit and a monostable circuit: a bistable circuit is not possible because the transformer cannot maintain a signal across the windings indefinitely.

Astable Circuits

One form of transistor blocking oscillator is illustrated in the circuit diagram of Fig. 13.12. The primary winding of the transformer is included in the collector circuit and the secondary winding in the base circuit. The black dots indicate points of similar instantaneous polarity and show the winding sense required for the required phase

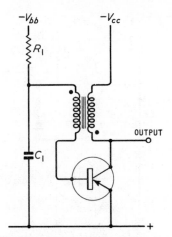

Fig. 13.12. An astable blocking oscillator

inversion in the transformer. Base bias is provided by the resistor R_1 and in this circuit R_1 is returned to a source of negative voltage: the transistor is a pnp type and this bias voltage ensures that the circuit will always revert to the state in which the transistor is conductive. As soon as conduction starts, however, positive feedback causes oscillation at the resonance frequency of the primary winding (assumed the larger of the windings). The design of the circuit is such that the oscillation amplitude builds up very rapidly and as it does so the capacitor C_1 is charged by the base current in the transistor. The voltage generated across C_1 by this current is such as to bias the base of the transistor positively with respect to the emitter. Such a polarity cuts a pnp transistor off, of course, and one of the aims in the design of blocking oscillators is to cut the transistor off in the first half-cycle of oscillation.

After cut off there is a period of relaxation in the circuit whilst C_1 discharges through R_1. As soon as the voltage across C_1 has fallen to a value at which the transistor starts to conduct, oscillation begins again and the cycle is repeated. Thus the circuit is astable (free-running) and takes simultaneous bursts of base and collector current at a repetition frequency governed by the time constant $R_1 C_1$, the duration of the burst being determined by the resonance frequency of the transformer primary winding. Provided the burst of collector current is large enough to bring the collector-emitter voltage down to nearly zero, pulses of substantially rectangular waveform can be obtained from the collector terminal as suggested in Fig. 13.12.

Very close coupling is desirable between the two windings of the

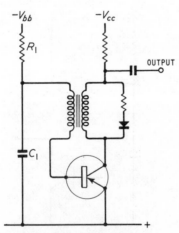

Fig. 13.13. An astable blocking oscillator with a diode to suppress overshoot

transformer to give a large degree of positive feedback and heavy damping of the windings is sometimes arranged to discourage ringing at the resonance frequency of the primary winding. A practical precaution which is usually desirable is to connect a diode across the primary winding as shown in Fig. 13.13. This has the effect of eliminating the large overshoot which would otherwise be generated across the primary winding when the collector current is suddenly cut off. Such overshoots could cause the collector-emitter voltage to exceed the safe rating for the transistor.

We can obtain an estimate of the natural frequency of the blocking oscillator in the following way. If the collector supply voltage is V_{cc} then a voltage with a peak value of nearly V_{cc} is generated across the primary winding when the transistor is suddenly turned on. We will assume the transformer to be a step-down type with a turns ratio of $n:1$. The voltage generated across the secondary winding when the transistor is turned on is then approximately V_{cc}/n. This is also the voltage generated across the capacitor C_1 because the base-emitter circuit of the blocking oscillator behaves in a manner similar to that of a diode detector and a capacitor in the circuit is charged up to the peak value of the applied signal. Initially the voltage across C_1 biases the base positively but C_1 is connected via R_1 to a source of negative voltage V_{bb}. Thus in the absence of the transistor the voltage across C_1 would initially fall to zero and then change sign, ultimately reaching the value V_{bb}, the curve following an exponential law. However, the moment the voltage reaches the cut-off value V_{be} (approximately -0.7 V for a pnp silicon transistor) the transistor begins to conduct. Oscillation then occurs and C_1 is again charged to V_{cc}/n volts. Thus only the initial part of the exponential change in voltage across C_1 is therefore achieved and the interval between successive bursts of oscillation (which, of course, determines the natural frequency of the blocking oscillator) is equal to the time taken for a voltage initially equal to $(V_{cc}/n + V_{bb})$ to fall by $V_{cc}/n + V_{be}$, the time constant being $R_1 C_1$.

Now the initial part of an exponential curve is nearly linear and has a slope given by $V_o/R_1 C_1$ where V_o is the initial voltage. Thus the fall in voltage in a time t is given by

$$\left(\frac{V_{cc}}{n} + V_{bb}\right) \cdot \frac{t}{R_1 C_1}$$

and if we equate this to $V_{cc}/n + V_{be}$ we can obtain an estimate of the natural frequency of the oscillator. We have

$$\frac{V_{cc}}{n} + V_{be} = \left(\frac{V_{cc}}{n} + V_{bb}\right) \cdot \frac{t}{R_1 C_1}$$

giving

$$\text{natural frequency} = \frac{1}{t}$$

$$= \frac{(1 + nV_{bb}/V_{cc})}{(1 + nV_{be}/V_{cc})} \cdot \frac{1}{R_1 C_1}$$

Control of Natural Frequency

If R_1 is returned to the collector supply voltage, $V_{bb} = V_{cc}$: if, in addition V_{cc}/n is large compared with V_{be} the natural frequency is given by

$$f = \frac{n+1}{R_1 C_1}$$

The natural frequency is now independent of the supply voltage and, for a given oscillator with a fixed value of n, can be controlled by adjustment of R_1 or C_1.

As a numerical example suppose a blocking oscillator is required to operate at 10 kHz. Such an oscillator is required in a 625-line television receiver, the synchronised frequency being 15 kHz. A typical value for the turns ratio of the transformer is 5:1 and re-arranging the above expression we have

$$R_1 C_1 = \frac{n+1}{f}$$

$$= \frac{6}{10^4} \, \text{s}$$

$$= 600 \, \mu\text{s}$$

If we wish the collector current to peak to 5 mA, the base current should preferably be not less than, say, 0·3 mA. R_1 must then be 80 kΩ to give such a base current with a 24-V supply. C_1 is then given by

$$C_1 = \frac{\text{time constant}}{R_1}$$

$$= \frac{600 \times 10^{-6}}{80 \times 10^3} \, \text{F}$$

$$= 0·008 \, \mu\text{F}$$

It is probably better, however, to have separate supplies for collector and base as shown in Fig. 13.13 and to control the natural frequency by adjustment of V_{bb} using fixed values of R_1 and C_1. This has the advantage that the natural frequency can be controlled by a potentiometer (as shown in Fig. 13.14) which need carry only direct current and can therefore be situated at some distance from the blocking oscillator itself.

Control of Pulse Duration

The duration of the burst of conduction is governed by the resonance frequency of the primary winding of the transformer, being approximately equal to half the period of oscillation. For example, if the winding resonates at 50 kHz the period is given by

$$T = \frac{1}{f} = \frac{1}{50 \times 10^3} \text{ s} = 20 \text{ } \mu\text{s}$$

and the burst of conduction is approximately 10 μs duration. Often blocking-oscillator transformers are designed to have a primary inductance and self-capacitance which will give the required duration. Alternatively precautions can be taken to keep self-capacitance to a minimum and the required duration can be obtained by adding a physical component of the correct capacitance across the primary winding.

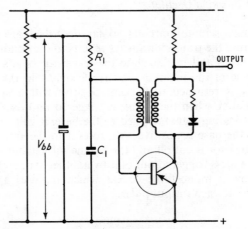

Fig. 13.14. *Control of natural frequency of a blocking oscillator by adjustment of V_{bb}*

Output Terminals of the Blocking Oscillator

The output can be taken from a blocking oscillator from the collector terminal as suggested in Fig. 13.12 but external loads added at this point could affect operation of the oscillator and it is better to include a resistor in the collector circuit as shown in Figs. 13.13 and 13.14. This gives positive-going output pulses (if the transistor is a pnp type). If negative-going output pulses are required an npn transistor can be used, the supply voltage being positive.

Synchronisation of Blocking Oscillators

Blocking oscillators can readily be synchronised at the frequency of any recurrent signal. Synchronisation is achieved by terminating the non-conductive period earlier than would occur naturally. Thus the natural frequency must be lower than the frequency of the sync pulses. To terminate the non-conductive period a negative-going signal can be applied to the base of the transistor (assumed pnp) or a positive-going signal can be applied to the collector. Alternatively if the transistor is an npn type, a positive-going signal is required at the base or a negative-going signal at the collector. The precise waveforms of the sync pulses are not important provided they have a steep leading edge.

Monostable Blocking Oscillator

The blocking oscillator circuits so far described are essentially astable because the transistor always reverts to the conductive state as C_1 discharges and in this state the transistor bursts into oscillation because of the inherent positive feedback in the circuit. If, however, R_1 is returned to a point positive with respect to the emitter potential, when the circuit is triggered into oscillation, e.g. by a negative-going signal applied to the base, it will revert to the state where the base is positive with respect to the emitter. In this state the transistor is cut off and the circuit will remain indefinitely in this state unless triggered out of it by another sync signal. This is therefore now a monostable circuit which requires a triggering signal to initiate each output pulse.

CHAPTER 14

Sawtooth Generators

INTRODUCTION

Sawtooth waves are extensively employed in electronic equipment. Two obvious applications are as sources of deflecting signals in oscilloscopes and television receivers but they are also used in radar and in test equipment.

Transistors can be used in many different types of sawtooth generator: in some they are used as switches and in others as linear amplifiers.

Usually sawtooth generators are classified as *free-running* (astable) which do not require external signals (but can readily be synchronised by such signals) and *driven* (monostable) which will not operate without external signals.

SIMPLE DISCHARGER CIRCUIT

It was mentioned in the previous chapter that the initial part of an exponential curve is almost linear and in many sawtooth generators this part of such a curve is used as the working stroke, a faster exponential change providing the rapid flyback which terminates this stroke.

One form of sawtooth generator operating on this principle is illustrated in the circuit of Fig. 14.1. TR1 has an input consisting of widely spaced negative-going pulses. In the intervals between the pulses TR1 is cut off and C_1 charges through R_1 to provide the working stroke of the output. When, however, TR1 is turned on by the input pulses TR1 takes collector current from C_1 and discharges it to produce the flyback.

This is essentially a driven circuit because there is no output in the

241

absence of input pulses. Increase in R_1 or C_1 reduces the rate of charge of C_1 with the result that a smaller voltage builds up across C_1 during the interval between successive pulses: in other words increase in the time constant R_1C_1 reduces the amplitude of the sawtooth output. Similarly reduction in the time constant increases the output amplitude but, if good linearity of forward stroke is

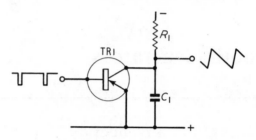

Fig. 14.1. A simple transistor discharger

required, the output should be kept small compared with the supply voltage to ensure that only the initial part of the exponential curve is used as output.

As an illustration the departure from linearity is 5 per cent if the output swing is restricted to 10 per cent of the supply voltage. The time constant of an RC combination is, by definition, the time taken for the voltage across a capacitor to rise to 63 per cent of the final value. The time taken for the voltage to rise to only 10 per cent of the final value is approximately 0·1 RC. Expressed differently the time constant of the network required in a simple discharger should be approximately 10 times the duration of the forward stroke. Thus if the forward stroke is 50 μs, a time constant of 500 μs is required.

There is, of course, an unlimited number of combinations of resistance and capacitance with this value of time constant but there are limitations on the values which can be successfully used and a decision can usually be made by considering the charging and discharging currents. The capacitor is charged via the resistor during the forward stroke and is discharged by the transistor during flyback. The ratio of average discharge current to average charge current is thus equal to the ratio of forward stroke period to flyback period and may be 10 or 20:1 in a practical circuit. Suppose the forward stroke is to occupy 50 μs and the flyback 3 μs. It is convenient to take the average discharge current as 20 mA for this is easily within the capability of even a small transistor. The sawtooth amplitude is to be 10 per cent of the supply voltage and we will assume an amplitude of 2 V. We can now calculate the capacitance required from the relationship

$$C = \frac{Q}{V} = \frac{it}{V}$$

Substituting for i, t and V for the discharge of C

$$C = \frac{20 \times 10^{-3} \times 3 \times 10^{-6}}{2} \; \text{F}$$

$$= 0 \cdot 03 \; \mu\text{F}$$

The average charging current is $3/50$ of 20 mA, i.e. $1 \cdot 2$ mA. The supply voltage is 20 V and the average value of voltage across the charging resistor during the forward stroke is 19 V. The current is $1 \cdot 2$ mA and the resistor value is clearly 16 kΩ approximately.

We can check the validity of this calculation from the time constant of the combination. This is $16 \times 10^3 \times 0 \cdot 03 \times 10^{-6}$, i.e. 480 μs, approximately 10 times the duration of the forward stroke as required.

The simple discharger circuit has a rectangular input signal from which it develops a sawtooth output signal of the same frequency. Mathematically the output waveform is the time integral of the input waveform and the circuit may therefore be described as an integrator.

The collector current of TR1 is, of course, turned on at regular intervals by the input signal and it follows that any circuit in which current flows as a series of recurrent bursts can be connected to a resistor and capacitor combination (an integrating circuit, in fact) such as R_1 and C_1 in Fig. 14.1 to produce a sawtooth output. For example, an RC combination can be added to the output circuit

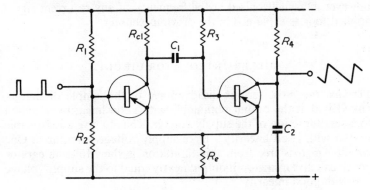

Fig. 14.2. *A driven multivibrator sawtooth generator*

of a multivibrator. Fig. 14.2 gives the circuit diagram of a monostable emitter-coupled multivibrator modified in this way to produce a sawtooth output. The time constant R_3C_1 determines the duration of each forward stroke of the sawtooth output whilst the time constant R_4C_2 determines the amplitude of the output.

Fig. 14.3 gives the circuit of an astable blocking oscillator used as a

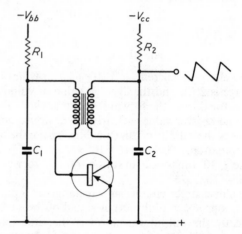

Fig. 14.3. Basic circuit of a blocking oscillator sawtooth generator

discharger. This is a free-running circuit, the natural frequency of which is determined by the time constant R_1C_1. The output amplitude is determined by both R_1C_1 and R_2C_2 and the ratio of these quantities should be kept constant if the output amplitude is required to be constant when the natural frequency is varied. This circuit can, however, be synchronised at the frequency of any recurrent signal applied to it as explained in the previous chapter.

MILLER-INTEGRATOR CIRCUIT

Perhaps the most serious disadvantage of the simple discharger circuit is that the output obtainable with good linearity is limited to a small fraction of the supply voltage. Thus to give a reasonable output with good linearity a large supply voltage is required. One circuit which is free from this limitation is the Miller integrator which can give a sawtooth output nearly equal to the supply voltage and with good linearity.

As shown in Fig. 14.4 the circuit is similar to that of the simple discharger but the capacitor C_1 is returned to the base instead of to the emitter. This introduces negative feedback and makes the operation of this circuit quite different from that of the simple discharger. For example the transistor in the Miller integrator is conductive throughout the cycle and is used as a class-A amplifier: in fact the transistor is used as an operational amplifier and this

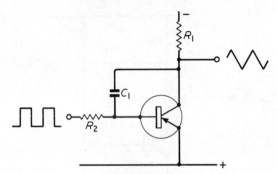

Fig. 14.4. Fundamental circuit of a transistor Miller integrator

approach is described on page 146. Suppose the transistor (assumed pnp) has suddenly been turned on by a negative-going pulse applied to the base. The immediate tendency is for the collector potential to approach the emitter potential as the collector current discharges C_1. As the collector potential changes, however, the change is transferred by C_1 to the base. The resulting feedback reduces the collector current and slows up the discharge of C_1. In fact the effective time constant is increased by the feedback from R_2C_1 to AR_2C_1 where A is the voltage gain of the transistor with R_1 as load. In addition to slowing up the discharge of C_1, the feedback exerts a powerful linearising effect and the sawtooth generated at the collector terminal is a much better approximation to the ideal than that obtained from the simple discharger circuit.

The reduction in the speed of the collector-voltage change is undesirable during the flyback period, which is usually required to be as short as possible, and a common technique is to include a second transistor in practical Miller-integrator sawtooth generators to interrupt the feedback circuit for the duration of the flyback period. One circuit of this type is illustrated in Fig. 14.5. The two transistors TR1 and TR2 are connected in series as a cascode but the input is applied to TR2 base, not TR1 base as in r.f. applications. The overall gain is approximately equal to that of a single transistor and there is also a phase reversal between the signals at TR1 base and TR2 collector. Thus the fundamental components R_1, R_2 and C_1 can be connected as shown, these corresponding with those of the previous diagram. During the period when the base is biased negatively by the input both transistors conduct and the circuit operates in the manner already described for Fig. 14.4, negative feedback slowing down the discharge of C_1 and linearising the

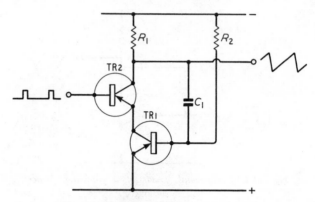

Fig. 14.5. *Transistor Miller-integrator circuit*
with fast flyback

sawtooth generated at TR2 collector. When, however, TR2 base is held positive by the input signal, TR2 is cut off. This removes negative feedback from the circuit and C_1 now charges through R_1 and the base-emitter junction of TR1 at a rate probably 40 or 50 times that which applies with feedback. In this way, therefore, a rapid flyback is obtained.

LINE OUTPUT STAGE FOR TELEVISION RECEIVER

The principles of the method of generating a sawtooth current commonly used in television receivers are quite different from those so far described. At frequencies of the order of 15 kHz the

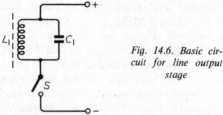

Fig. 14.6. *Basic cir-*
cuit for line output
stage

scanning coils are predominantly inductive and are represented by L_1 in Fig. 14.6. The parallel capacitor is chosen to resonate with L_1 at a frequency such that the period is double the required fly-back time.

When S is closed (t_1 in Fig. 14.7) it connects L_1 across the supply. Current in L_1, initially zero, starts to grow and rises linearly with

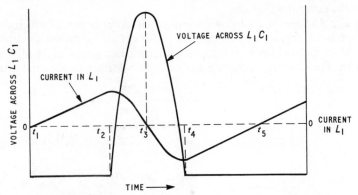

Fig. 14.7. *Voltage and current waveforms for the circuit of Fig. 14.6*

time: the voltage across an inductor is related to the rate of change of current in it according to the expression

$$V = -L\frac{di}{dt}$$

V in this circuit is the supply voltage and is constant: L_1 is also constant. Thus di/dt is constant, i.e. the current changes linearly with time as required for the working stroke of the scanning system. When the working stroke is completed, S is opened (t_2), isolating L_1C_1 from the supply. L_1C_1 now begins to oscillate freely and the following voltage and current changes occur.

Current continues to flow in L_1 but is now taken from C_1 instead of from the supply. This discharges C_1 reducing the voltage across L_1 and thus reducing the rate of rise of current in L_1. When C_1 is completely discharged, the current in L_1 has zero rate of rise, being now at a maximum. As current continues to flow C_1 begins to charge up with the opposite polarity. This causes current in L_1 to decrease and it falls to zero and builds up in the opposite direction. At the instant (t_3) when the current is zero, the voltage across C_1 is a maximum. As soon as the current has reached a value equal to the previous maximum, S is closed again (t_4). At this instant the voltage across C_1 is equal to the supply voltage. Current now flows from L_1 into the supply and because the supply voltage is constant the current falls linearly with time to zero (t_5). This completes the first half of the working stroke and current now begins to increase linearly, as described initially, to complete the second half of the stroke. The current and voltage waveforms are given in Fig. 14.7: the voltage waveform is of course proportional to the rate of change of the current waveform.

Fig. 14.8 shows a simplified realisation of the circuit in which the switch S is replaced by an npn transistor driven by pulses, e.g. from a blocking oscillator possibly via intermediate stages of amplification. The scanning coils are fed via a transformer to eliminate the

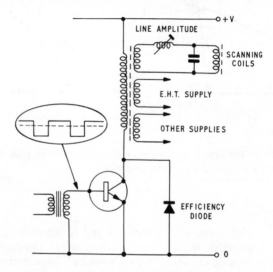

Fig. 14.8. Simplified circuit diagram for a line output stage

direct component of the transistor collector current which would otherwise give an undesirable static deflection of the beam. The pulses hold the transistor conductive for the second half of the working stroke during which current grows linearly in the scanning coils, the energy being taken from the supply. At the end of the flyback period the collector voltage swings negatively, as shown in Fig. 14.7, and drives the collector negative with respect to the emitter. The collector-base junction (which is reverse-biased during normal amplification) now becomes forward-biased and provides a low-resistance path between collector and the negative supply via the secondary winding of the driver transformer. Thus the primary of the line output transformer is effectively connected across the supply and the current in the scanning coils (now in the opposite direction to that before flyback) falls linearly to zero: in this way the energy supplied to the scanning coils during the second half of the previous working stroke is now returned to the supply. Sometimes, as shown in Fig. 14.8, an additional diode, known as an efficiency diode, is connected in parallel with the line output transistor to reduce still further the resistance of the path by which energy is returned to the supply. Provision is required for adjustment of the sawtooth amplitude to give control of picture width: a variable inductor in series with the scanning coils can be used. The line output stage is also used as source of e.h.t. supply, boosted collector supply voltage for the line output stage itself and supply voltages for other transistors in the receiver.

CHAPTER 15

Digital Circuits

INTRODUCTION

As mentioned in the previous two chapters, transistors in multi-vibrators and other pulse circuits are used as switches, i.e. except for the brief periods during which the transistors are changing state they are either *on* (taking considerable current) or *off* (non-conductive). For most of the time therefore the collector (or drain) voltage has one of two possible values namely a low value near emitter or source potential (when the transistor is on) and a high value near the supply potential (when the transistor is off).

A vast range of circuits has been developed during the past decade or two in which diodes and transistors are used as switches and in which the signal paths have at all times one or other of two possible voltage levels. Such circuits are used to perform mathematical and logical operations on signals in computers and similar equipment. Circuits used in this manner are known as digital or logic circuits and the principal types are described briefly in this chapter.

LOGIC LEVELS

The two significant values of voltage on the signal-carrying lines are referred to as logic level 0 and logic level 1.* If level 1 is more positive than level 0 the circuit is said to use the *positive logic convention* and if level 1 is more negative than level 0 the circuit is said to use the *negative logic convention*. The distinction is important because circuits can behave differently according to the logic convention

---

* The two levels could alternatively be values of current or more generally of air pressure in pneumatic systems or fluid pressure in hydraulic systems.

chosen. This is illustrated below but it is worth stressing now that the logic convention chosen should always be stated or implicit on diagrams of logic circuits.

BINARY SCALE

The advantage of labelling the two significant voltage levels 0 and 1 is that it simplifies the process by which logic circuits are able to carry out mathematical and other operations. Arithmetical operations, for example, can be performed by using the binary scale of numbers which has only two digits 0 and 1.

Conventional counting uses the scale of 10 (the decimal scale) and in numbers the digits are arranged according to the power of 10 they represent. For example the number 4721 (four thousand, seven hundred and twenty-one) means, if written out in full:

$$4 \times 10^3 + 7 \times 10^2 + 2 \times 10^1 + 1 \times 10^0$$

$$\text{i.e.} \quad 4000 + 700 + 20 + 1$$

$$= 4721$$

Similarly in the binary scale of counting, the digits (0 or 1) in a number are arranged according to the power of 2 they represent. For example the binary number 110101 means, if written out in full:

$$1 \times 2^5 + 1 \times 2^4 + 0 \times 2^3 + 1 \times 2^2 + 0 \times 2^1 + 1 \times 2^0$$

$$\text{i.e.} \quad 32 + 16 + 0 + 4 + 0 + 1$$

$$= 53$$

The first nine numbers in the binary scale are as follows:

binary number	decimal equivalent
1	1
10	2
11	3
100	4
101	5
110	6
111	7
1000	8
1001	9

It is common practice, however, with very large numbers to translate each digit of the decimal number separately into the binary scale. This is known as the *binary-coded decimal* system and in it the number 4 721 would be coded as follows:

$$100 : 111 : 10 : 1$$

$$\text{i.e.} \quad 4 \ : \ 7 \ : \ 2 : 1$$

This system has the advantage that after a little experience binary coded numbers can be translated into decimal form on inspection. It also simplifies the design of equipment for coding decimal numbers into binary form and for decoding and displaying binary-coded numbers in decimal form.

LOGIC GATES

A simple logic circuit employing two diodes is illustrated in Fig. 15.1. Input *A* is at all times at +10 V or 0 V, these being the standard voltage levels chosen for use in this circuit. Input *B* is similarly at

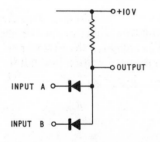

Fig. 15.1. Simple diode gate circuit. It can be an AND gate or an OR gate depending on the logic convention adopted

+ 10 V or 0 V. If either input is at 0 V the associated diode conducts and, if the forward resistance of the diode is neglected, 0 V appears at the output. A diode with an input at + 10 V does not conduct and

Table 15.1

Voltages in the circuit of Fig. 15.1

Input volts		Output volts
A	*B*	
0	0	0
0	+ 10	0
+ 10	0	0
+ 10	+ 10	+ 10

the output is isolated from this input. For two inputs there are four possible combinations of input voltage. They are shown in Table 15.1 together with the value of the output voltage. This table shows that the output is 0 V if either or both of the inputs is at 0 V and that the only way to obtain $+10$ V at the output is for both of the inputs to be at $+10$ V.

If the positive logic convention is used $+10$ V is logic level 1 and 0 V is logic level 0. If Table 15.1 is repeated in terms of logic levels,

Table 15.2

Truth table for an AND gate

Input		Output
A	B	
0	0	0
0	1	0
1	0	0
1	1	1

the result is as shown in Table 15.2. Such a table is known as a *truth table* and it shows that a logic 1 is obtained at the output of the circuit of Fig. 15.1 only when input *A and* input *B* have a logic 1 signal. A circuit such as this which requires a logic 1 signal at all the inputs to give a logic 1 signal at the output is known as an AND gate.

Suppose negative logic is used in the circuit of Fig. 15.1. Now $+10$ V represents logic level 0 and 0 V represents logic level 1.

Table 15.3

Truth table for an OR gate

Input		Output
A	B	
0	0	0
0	1	1
1	0	1
1	1	1

The truth table now has the form shown in Table 15.3. A logic 1 is obtained at the output of the gate when either input *A or* input *B* has a logic 1 signal. A gate which requires a logic 1 signal at any one input to give a logic 1 at the output is known as an OR gate.

Thus the circuit illustrated in Fig. 15.1 can behave as an AND gate or an OR gate depending on the logic convention used. In general it is not possible to state the nature of a logic circuit until the logic convention to be used with it is known.

Suppose a common-emitter amplifier is added after the diode gate as shown in Fig. 15.2. The transistor circuit is so designed that

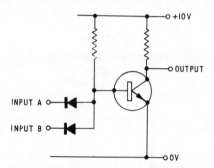

Fig. 15.2. An inverter stage following the diode gate gives a circuit which can be a NAND gate or a NOR gate depending on the logic convention adopted

its collector voltage is always at one or other of the two chosen voltage levels. The signal inversion introduced by the amplifier gives yet another type of behaviour and the relationship between the inputs and output, for positive logic, is as given in Table 15.4.

Table 15.4

Truth table for a NAND gate

Input		Output
A	B	
0	0	1
0	1	1
1	0	1
1	1	0

This shows that the only way to obtain a logic 0 at the output is for all the input signals to be at logic 1. This behaviour is not surprisingly the inverse of that of the AND gate: this circuit, for positive logic, is therefore known as a NOT-AND or more simply a NAND gate.

For negative logic the behaviour of the circuit of Fig. 15.2 is as illustrated in the truth table of Table 15.5. When any of the input signals is at logic 1 the output is at logic 0. The only way to obtain a logic 1 at the output is for all the inputs to be at logic 0. This

Table 15.5

Truth table for a NOR gate

Input		Output
A	*B*	
0	0	1
0	1	0
1	0	0
1	1	0

behaviour is the inverse of that of the OR gate and this circuit, for negative logic, is known as a NOT-OR or NOR gate.

Gate Symbols

For simplicity only two inputs are shown in Figs. 15.1 and 15.2 but there can, of course, be more. The block symbols for the four basic types of gate so far introduced are shown in Fig. 15.3 and here three

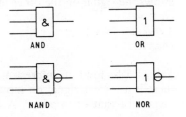

Fig. 15.3. Block (or logic) symbols for the four basic types of gate

inputs are shown. The circle at the output of the NOR and NAND gates indicates the inversion of the output relative to that of the OR and AND gates.

Integrated Circuit Gates

Gate circuits can be constructed of discrete components but normally they are in the form of integrated circuits. For example a single i.c. may contain three 3-input gates. To use such an i.c. in an equipment it is not necessary to know details of the circuitry of the device. All the designer needs to know to be able to use the i.c. successfully are details of input and output signal levels, polarities, impedances and supply voltages. In preparing diagrams of computers and computer-like equipments the gates and other functional units are represented by block symbols such as those given for gates in Fig. 15.3. To help in the layout of printing wiring cards and in maintenance the inputs, outputs and supply points of the gates can be identified in the block diagram by giving the pin numbers of the i.c.s. Block diagrams of logic equipment are usually known as logic diagrams.

To illustrate the versatility of logic gates a number of applications will now be considered.

Gates as Switches

Table 15.2 shows that when input A is at logic 0, the output is also at logic 0 (irrespective of the signal on input B) whereas if input A is at logic 1 the output signal is the same as the signal on input B. Thus an AND gate can be used as a switch, a logic 1 signal on input A allowing the signal on input B to pass through the gate, a logic 0 signal on input A blocking the signal on input B. An example of such a use of an AND gate is given later.

An OR gate can be used similarly but here a logic 0 on one input allows the signal on the other input to pass through the gate.

Inverters

Table 15.4 shows an interesting property of a NAND gate. When there is a logic 1 signal on input A the output from the gate is the inverse of that on input B. A NAND gate is frequently so used,

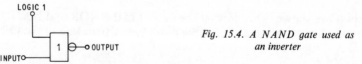

Fig. 15.4. A NAND gate used as an inverter

input A being connected permanently to a source of logic 1 voltage as shown in Fig. 15.4. For some types of NAND gate i.c.s it is sufficient simply to leave input A unconnected: this has the same effect as connecting it to a logic 1 source.

A NOR gate can also be used as an inverter but one input must be connected to a source of logic 0 voltage to obtain inversion of the signal on the other input.

Inhibiting Input

Reference to Table 15.5 shows that if one input is at logic 1, the output of the gate is prevented from taking up the logic 1 state irrespective of the signal on the other input. This ability of an input of a device to prevent the output taking up the logic 1 state is sometimes exploited in logic circuitry and an input so used is then known as an inhibiting input: it is indicated on logic diagrams by a short stroke across the signal line to that input.

Exclusive-OR Gate

Combinations of gates can be used to perform desired operations. For example suppose a circuit is required to give a logic 1 output when either of two inputs is at logic 1 but not when both are at logic 1. Such a circuit is known as an *exclusive-OR* gate.

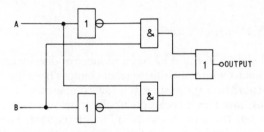

Fig. 15.5. One possible logic circuit for an exclusive-OR gate

There are a number of possible circuits and one is given in Fig. 15.5. That this circuit gives the required performance can be checked from the truth table of Table 15.6 or by the use of Boolean algebra and this second method is the one normally used in designing logic

Table 15.6

Truth table for an exclusive-OR gate

Input		Output of inverter A	Output of inverter B	Output of AND gate A	Output of AND gate B	Output of OR gate
A	B					
0	0	1	1	0	0	0
0	1	1	0	1	0	1
1	0	0	1	0	1	1
1	1	0	0	0	0	0

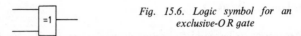

*Fig. 15.6. Logic symbol for an
exclusive-OR gate*

gate circuits. The exclusive-OR gate is usually represented by the
logic symbol of Fig. 15.6.

Equivalence Element

It is sometimes necessary to compare two binary digits and to give
an indication when they are the same. A circuit used for this purpose
is known as an *equivalence element* or a *comparator* and is required
to give a logic 1 output when the two inputs are both at logic 1 or
both at logic 0. The required output is the inverse of that of the
exclusive-OR gate and can be obtained from a combination of
gates similar to that shown in Fig. 15.5 but with the OR gate replaced
by a NOR gate.

Practical NAND Gate Circuit

The circuit shown in Fig. 15.2 has a number of disadvantages. One
is that the speed with which the output changes from the 0 state to
the 1 state (positive logic assumed) is low compared with that of the
reverse transition. This is because a bipolar transistor can be switched
on quickly but the time of switch off is appreciable. The delay in
switch off slows up the rate at which logic operations can be carried

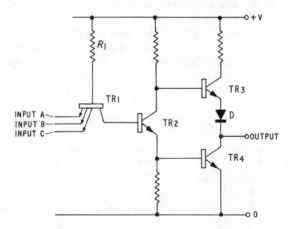

Fig. 15.7. Circuit diagram of an i.c. NAND gate

out. The effect can be avoided by using an output stage with two transistors in series, one of which is switched on to perform one transition, the other being switched on to give the opposite transition. The push-pull output circuit shown in Fig. 15.7 fulfils these conditions. The input diodes of Fig. 15.2 have been replaced by a common-base stage in which the transistor has a number of emitters, each providing one gate input. It is easy to fabricate such transistors using i.c. techniques but if discrete components are used one transistor would be required for each emitter. The second transistor in Fig. 15.7 is a phase splitter driving the push-pull output stage. Diode D provides base bias for TR3.

For positive logic this circuit is a NAND gate. To confirm this, assume the output to be at logic 0. This requires TR3 to be off and TR4 on. This, in turn, requires TR2 to be on. TR2 must therefore have a high base potential which requires TR1 to be off. The only way for TR1 to be off is for all the emitters to be at positive supply potential, i.e. at logic 1. By definition a gate which requires all inputs to be at logic 1 to give a logic 0 output is a NAND gate.

Fan In

In circuits containing numerous gates it may happen that several gates feed one particular gate. Several output circuits are then connected in parallel across one input circuit and it is essential that this loading should not affect the voltage level at the junction point, no matter whether this is the logic 1 or logic 0 voltage. There is normally a tolerance on the voltage levels but if too many circuits are connected together the junction voltage may be outside the tolerance and normal circuit behaviour becomes impossible. The greatest number of outputs which may be connected to a gate input whilst still permitting normal behaviour is known as the fan in. This may in practice be as high as 10 but in the gate circuits and symbols given earlier the number of inputs was shown for simplicity as only 2 or 3.

Fan Out

A junction between input and output circuits also occurs when one gate is required to feed a number of others: one output circuit is then connected to several input circuits. Current is required to operate an input circuit using bipolar transistors. For example in Fig. 15.7, to put a logic 0 signal on an emitter of TR1, the emitter potential must be lowered to that of the supply negative voltage (positive logic is assumed). An emitter current (determined by R_1 and the supply voltage) of, for example, 1·5 mA then flows in the input circuit and the output circuit feeding it. The conductive transistor in the output circuit must be capable of passing this current without undue rise in the voltage across it. Any significant rise may cause this voltage to fall outside the range recognised as a logic 0 signal. The maximum collector current through the output transistor may be, say 15 mA, which enables up to 10 gate circuits to be fed satisfactorily. This maximum number is known as the fan out of the circuit.

Distributed AND Connection

It is sometimes possible, when the outputs of two or more gates are paralleled, to achieve a logic AND function at the connection point without including a gate circuit at the point for the purpose. As an example suppose in a positive logic system the output circuit of a number of gates consists simply of an npn transistor without a collector load resistor: two such output stages are shown in Fig. 15.8. If the collectors are connected to the positive supply terminal

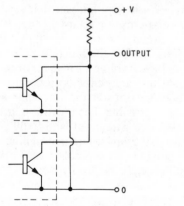

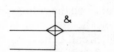

Fig. 15.9. Logic symbol for a distribution AND connection

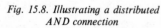

Fig. 15.8. Illustrating a distributed
AND connection

via a common load resistor then the only way in which a logic 1 signal can be obtained at the common connection point is by cutting off all the transistors. Thus all the output transistors must be in a logic 1 state to give a logic 1 at the common output point. An AND function is thus obtained at the interconnection point irrespective of the nature of the individual gates shown by the dashed lines in the diagram.

An AND function so obtained is known as a distributed or wired AND: it is represented in logic diagrams by the symbol given in Fig. 15.9.

Distributed OR Connection

The circuit of Fig. 15.8 can also be used to give an effective OR function at the common connection point. One simple way of achieving this is to use negative logic. When any of the output transistors is conductive the potential of the common point

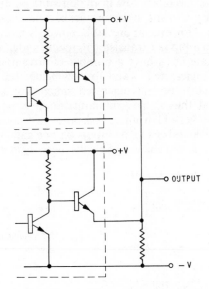

Fig. 15.10. Illustrating a distributed OR connection

approaches that of supply negative, i.e. when any of the transistors is in the 1 state, the output is in the 1 state: this is the OR function.

Alternatively a distributed OR connection can be obtained by use of different circuitry and one possible circuit is shown in Fig. 15.10. The output stages are emitter followers and the emitter

Fig. 15.11. Logic symbol for a distributed OR connection

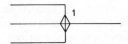

connections are brought out to an external load resistor which is returned to supply negative. We now obtain a logic 1 output signal (positive logic assumed) when any one output stage is at logic 1. The distributed or wired OR connection is represented by the logic symbol shown in Fig. 15.11.

Insulated-gate f.e.t. i.c.s as Logic Gates

I.C.s using bipolar transistor techniques as so far discussed in this chapter are extensively used. They have the advantages that they operate satisfactorily with low collector voltages, switch-on times are low and they have a high current-handling capacity. Dissipation in the collector circuits is low in the off state because the collector current is nearly zero and in the on state because the collector voltage is nearly zero. However appreciable power is required in the base circuit to keep a bipolar transistor in the on state.

Insulated-gate f.e.t.s have a number of advantages over bipolar transistors in logic-gate i.c.s and are now being used. One advantage is that f.e.t.s can be manufactured much smaller than bipolar transistors and thus greater miniaturisation is possible. Secondly because of the very high input resistance of f.e.t.s it is possible to parallel a large number of gate inputs on one gate output, i.e. a very high fan out can be achieved. Thirdly there is no dissipation in an

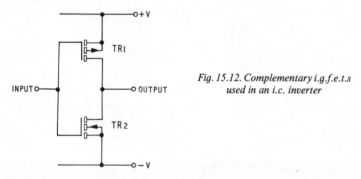

Fig. 15.12. Complementary i.g.f.e.t.s
used in an i.c. inverter

f.e.t. in the on or off state. A disadvantage of the f.e.t. is that it can easily be damaged by excess voltages and its current-handling capacity is less than that of bipolar transistors.

An example of an inverter circuit using i.g.f.e.t.s is given in Fig. 15.12. It could hardly be simpler and no resistors (to dissipate power) are necessary. Two complementary i.g.f.e.t.s are connected in series across the supply and the gates are connected to the input signal. A logic 1 signal (positive logic assumed) cuts TR1 off and turns TR2 on so that a logic 0 signal appears at the output. A logic 0 input turns TR1 on and cuts TR2 off so that a logic 1 output is obtained.

In general the output of a gate circuit may change when the inputs are removed: in other words gates have no memory. Logic operations must frequently be carried out in sequence and thus there is a need for a device with a memory, i.e. which can store logic information. Bistable multivibrators (usually abbreviated simply to bistables) are commonly used for this purpose. It was pointed out in the description of the multivibrators that the two transistors alternate between the on and off states. Thus the collector potential is either near that of the supply positive or the supply negative and if the supply potentials are taken as logic levels we can say that at any moment one collector voltage will be at logic 1 level whilst the other is at logic 0 level. After a change of state the collector potentials are reversed. The collectors provide the outputs of the bistable circuits and these are always in complementary states: the outputs are denoted by Q and $\bar{Q}$.

RS *Bistable*

A simple form of bistable circuit is illustrated in Fig. 15.13. Outputs are taken from the collectors of the transistors and two inputs known as S (set) and R (reset) are connected to the bases via diodes arranged to conduct negative-going signals. A negative-going signal applied to the S input cuts off TR1 (and hence turns TR2 on) and the circuit will now remain in this state indefinitely if need be, so storing information. To remove the information and restore the circuit to the state it had originally, a negative-going signal is applied to the R terminal which cuts off TR2 and turns TR1 on.

If negative-going inputs are applied to the R and S terminals simultaneously there is no way of knowing what the resultant state of the circuit will be because it is not natural, in a bistable such as this, to have such signals on both bases at the same instant. *RS* bistables are therefore never used in circumstances where simultaneous negative-going R and S inputs are possible. Positive-going signals applied to the R and S inputs have no effect on the circuit which remains in its previous state.

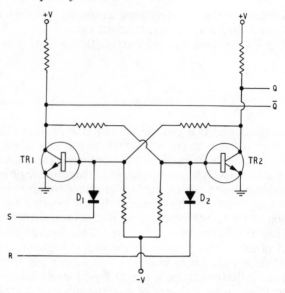

Fig. 15.13. Basic RS *bistable circuit*

Thus for three of the four possible combinations of *R* and *S* inputs, the resultant state of the bistable is predictable: for the fourth combination the output is indeterminate. This behaviour is summarised in the following table in which a logic 1 signal is taken to be a change to the more positive voltage level and a logic 0 signal is a change to the less positive level.

The *RS* bistable can also be used in a clocked mode. For this purpose a diode gate (see page 221) is included and a regular train of pulses, known as clock pulses, is fed to the gate. The circuit is so arranged that the clock pulses have no effect on the circuit until an input is applied to the *R* or *S* terminals, after which the next clock

Table 15.7

Truth table for an *RS* bistable

R input	S input	Q output
0	0	indeterminate
0	1	1
1	0	0
1	1	no change

pulse initiates the change of state of the bistable. The behaviour of the bistable is still as indicated in Table 15.7 but the third column should be interpreted as giving the logic level of the Q output after receipt of a clock pulse.

JK *Bistable*

The JK bistable may be regarded as an improved form of the RS bistable in which there is no indeterminate state. The improvement is achieved by arranging for the signal inputs to operate on the gating diodes D1 and D2 to which the clock pulses are applied.

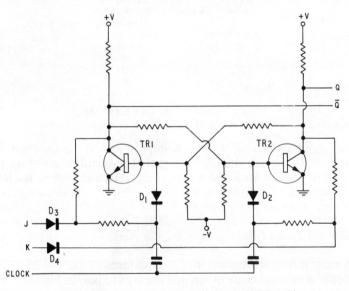

Fig. 15.14. Circuit diagram of a typical clocked JK bistable

A typical circuit diagram is given in Fig. 15.14. D1 and D2 are biased by the difference between the collector and base potentials of the associated transistor. If the transistor is on, there is little potential difference and the diode can conduct negative-going clock signals to the base to cut the transistor off. When the transistor is off, the considerable difference between collector and base potentials reverse-biases D1 or D2 so that clock signals cannot reach the base. A positive signal on the J input forward biases D3 and reverse-biases the clock-pulse diode D1 and thus prevents TR1 being cut off by the clock pulses. Similarly a positive signal on the K input forward biases D4 and reverse-biases D2, so preventing TR2 being cut off by the clock pulses. If *J* and *K* inputs are made positive simultaneously clock pulses are prevented from reaching either base and have therefore no effect on the bistable which remains in its former state. Negative-going signals on the *J* and *K* inputs are blocked by D3 and D4 so that clock pulses operate on the bistable and the outputs alternate between supply positive and negative at the clock frequency. The behaviour of the bistable is thus predictable for all

Table 15.8

Truth table for a *JK* bistable

J input	K input	Q output after receipt of a clock pulse
0	0	no effect
0	1	0
1	0	1
1	1	C

C = complement of the state before the clock pulse

four combinations of *J* and *K* inputs and is summarised in Table 15.8 in which the logic 1 signal is regarded as a change to the more positive level and a logic 0 signal as a change to the less positive level.

Clear and Preset Inputs

RS and *JK* bistables commonly have two further inputs which override the signal and clock inputs and are used to put the bistable into a desired state. These are the *clear* and *preset* inputs: they are shown in the block symbol of Fig. 15.15 which is used to represent the bistable in logic diagrams. The information to be stored is fed into the circuit via the signal inputs, the convention being that a logic 1 signal on the *J* or *S* input puts *Q* to logic 1 (and hence $\bar{Q}$ to logic 0) on receipt of a clock pulse. A logic 1 signal on input *K* or *R* has the opposite effect. A logic 0 signal on the preset input puts *Q* to logic 1 and a logic 0 signal on the clear input has the opposite effect.

This convention means that a logic 1 input to the upper square of the block symbol causes the output from that square to go to logic 1 on a clock pulse. Clearly, therefore, this square cannot represent one transistor of the bistable circuit because the signal inversion of the common-emitter amplifier is missing. The square represents the input circuit of one transistor and the output circuit of the other.

Sometimes the *JK* bistable has two parts known as the master

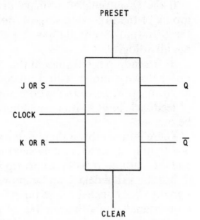

Fig. 15.15. Logic symbol for a bistable

and slave. On the positive-going edge of the clock pulse the information on inputs *J* and *K* is transferred to the master section and on the negative-going edge the information is transferred to the outputs Q and $\bar{Q}$.

Binary Dividers and Counters

Bistables give one alternation in output signal (at Q and $\bar{Q}$) for every two clock input signals. They can thus be used as binary dividers and Fig. 15.16 shows a cascade of bistables so used, each Q output

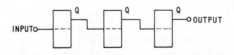

Fig. 15.16. A logic diagram of a simple up counter

being connected to the clock input of the following stage. When the clock pulses are started the pattern of the Q outputs follows the binary scale of numbers. Thus if all the Q outputs are initially at logic 0, then after seven clock pulses the three Q outputs will be at logic 1: 111 is the binary equivalent of seven. The next clock signal causes all Q outputs to go to logic 0 and the counting process begins again. The Q output of the final stage changes from logic 0 to logic 1 (and from logic 1 to logic 0) once for every eight clock inputs: thus the circuit has a division or count ratio of 8. A binary counter of this type is known as an up counter.

If the $\bar{Q}$ outputs of a binary divider are connected to the clock inputs of the following stage, a down counter is obtained, i.e. if all the $\bar{Q}$ outputs are initially at logic 1, after seven input signals they are all at logic 0.

By cascading n bistables in this manner it is possible to realise an up or down counter with a ratio of 2^n. Such a device is of limited application because the ratio can only be a power of 2 but by use of feedback any desired count ratio can be obtained as explained below.

There is inevitably a slight delay in a bistable: in other words there is a significant time lag between the leading edge of an input signal and the leading edge of the corresponding output signal. In a cascade of bistables this delay can be appreciable and it enables the output signal to be returned to the input of the cascade and accepted as an input signal. Thus the number of input signals required to deliver one output signal from a cascade of n bistables is reduced from 2^n to $(2^n - 1)$. An example of a counter using feedback in this way is

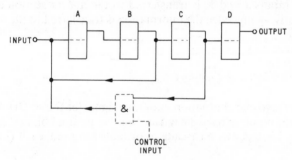

Fig. 15.17. Binary divider with feedback controlled by an AND gate

given in Fig. 15.17. Bistables A and B have together a count ratio of 4: this is reduced to 3 by the feedback loop embracing them. Thus the count ratio of bistables A, B and C is 6 which is reduced to 5 by the feedback loop embracing these three bistables. The final bistable (D) brings the overall count ratio to 10.

In certain applications of counters it is desirable to be able to alter the ratio. This can be done by including an AND gate in the feedback loop as shown in dotted lines in Fig. 15.17. Here by changing the control signal from logic 1 to logic 0 the division ratio can be altered from 10 to 12.

Waveform Generators Using Counting Techniques

Digital equipment commonly requires a number of control pulses of particular duration and timings. A method of generating such

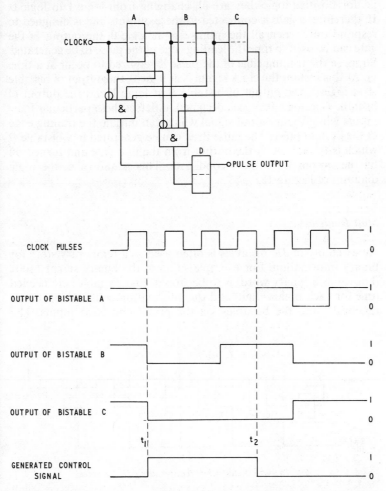

Fig. 15.18. Illustrating the digital method of generating a pulse

pulses is to use a train of bistables operated from a master or clock pulse source. The signals obtained at the outputs of the various stages of the bistable train have the form shown in Fig. 15.18. This is a simplified diagram and idealised in that the propagation time is ignored but will serve to illustrate the principle of this type of waveform generator.

Suppose the pulse to be generated is required to start at t_1. At this instant all the pulse trains shown have a negative-going edge, i.e. for positive logic they are all changing from logic 1 to logic 0. If, therefore, a gate is connected to these outputs and is designed to respond only when all the inputs are at logic 0, the output of the gate can be used to time the leading edge of the pulse to be generated. Suppose the trailing edge of the pulse is required to occur at a time t_2. At this instant the clock signal is at logic 0, the output of bistable A at logic 1, the output of bistable B at logic 1 and the output of bistable C at logic 0. A gate designed to detect these particular logic inputs will give an output signal which can initiate the trailing edge of the desired pulse. The pulse itself can be generated in a bistable D which is turned on by the output from the first gate and turned off by the output from the second gate. This is shown in the logic diagram of Fig. 15.18.

Shift Registers

An assembly of JK bistables is often used as a temporary store for binary information. For example suppose the binary signal 11001 (known as a binary word) is to be stored. Five bistables are needed (one for each binary digit) and the information could be fed simultaneously into the bistables via the preset and clear inputs. The

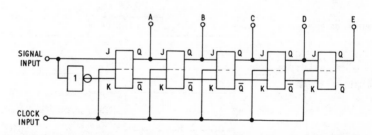

Fig. 15.19. A 5-stage shift register using JK bistables

information can now be stored for the necessary period and then simultaneously read out by operation of the clock inputs. This is the parallel method of feeding in and reading the information.

Alternatively the bistables may be connected as a counter as shown in Fig. 15.19 and the information can be fed into the input of the first bistable one digit at a time. The clock inputs of all bistables are commoned and the clock rate must be the same as the rate of receipt of the digits in the incoming information (the bit rate). On the first clock pulse the first binary digit (1) is transferred to the first bistable, its Q output registering logic 1. On the second clock pulse the first digit is transferred to the second bistable and the second digit (also 1) is transferred to the first bistable. Thus the process continues, the stored information moving to the right in the cascade of bistables until, after the fifth clock pulse, the whole binary word has entered the register. The Q outputs, in order, now read 11001. At this point the clock pulses can be stopped and the word stored for as long as necessary. When the word is required to be read out, the clock pulses are restarted and the stored information again moves to the right at clock rate and can be read at the Q output of the fifth bistable. This is the serial method of storing and reading information in a shift register.

It is possible to enter the information into the register by parallel methods and to read it out serially: the converse is also possible. Thus there are serial/parallel and parallel/serial shift registers.

CHAPTER 16

Further Applications of Transistors and other Semiconductor Devices

INTRODUCTION

In this chapter we shall describe a number of miscellaneous applications of junction diodes and transistors which do not properly belong in earlier chapters.

SIMPLE VOLTAGE-STABILISING CIRCUIT

It was pointed out in Chapter 1 that junction diodes could be used as a source of stable voltage and to begin this chapter we shall describe a simple circuit suitable for use where only a small current is required from the stable voltage source.

An example of such a requirement may occur in a transistorised car radio where the supply for the oscillator transistor may require stabilising against changes in car battery voltage to secure good frequency stability and hence stable tuning. A suitable circuit is

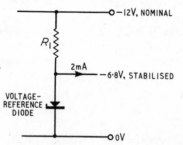

Fig. 16.1. Simple junction-diode voltage-stabilising circuit suitable for small currents

272

illustrated in Fig. 16.1. The junction diode must have a break-down voltage equal to the value of the stabilised voltage required and the value of R_1 must be chosen to give an operating point on the nearly-vertical part of the diode characteristic (Fig. 16.2). The

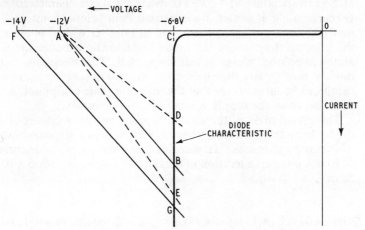

Fig. 16.2. *Illustrating the operation of a simple voltage-stabilising circuit*

diagram represents conditions in the circuit. The load line AB meets the axis at A at a voltage equal to the supply voltage, say 12 V. AB meets the diode characteristic at B and this point corresponds with the stabilised output voltage. OC represents the voltage drop across the diode and AC the voltage drop across R_1. In the chosen example the stabilised voltage is 6·8 V.

The slope of AB corresponds to the resistance R_1 and this clearly may vary within limits (as suggested by the dotted lines AD and AE) without much effect on the value of the stabilised voltage but it is preferable to choose a value for R_1 which keeps the dissipation in the diode well within the maximum value prescribed by the makers. For example if the maximum dissipation is 50 mW we can choose to dissipate half of this, 25 mW, at a battery voltage of 12 V. The voltage across the diode is 6·8 V and the diode current must be 25/6·8, i.e. 3·7 mA. This current is supplied via R_1 together with the current for the load (the oscillator transistor). If the load current is 2 mA, the total current in R_1 is 5·7 mA. The voltage across R_1 is 5·2 V and the required value of R_1 is given by

$$R_1 = \frac{5 \cdot 2}{5 \cdot 7 \times 10^{-3}}$$

$$= 900 \ \Omega \text{ approximately}$$

The supply voltage may easily rise to 14 V when the car dynamo is running. The effect such a voltage rise has on the stabilised voltage is illustrated by the load line *FG* which is parallel to *AB* (thus representing the same value of resistance R_1) but meets the axis at *F* corresponding to 14 V. *FG* meets the diode characteristic at *G* representing a greater diode current than before (point *B*). The new stabilised voltage corresponds to point *G* which, because of the extreme steepness of the diode breakdown characteristic is at almost the same voltage as before (point *B*). The dissipation in the diode is now greater than before the increase in supply voltage and care must be taken to see that the maximum safe dissipation is not exceeded when the supply voltage is at its maximum.

The effectiveness of the circuit depends on the steepness of the diode characteristic which is usually expressed as a slope resistance. This may be as low as 5 Ω, showing that a change of diode current of 10 mA gives an alteration in breakdown voltage of only $5 \times 10 = 50$ mV.

VOLTAGE-STABILISING CIRCUITS INCLUDING TRANSISTORS AND VOLTAGE-REFERENCE DIODES

The maximum current which can be drawn from a simple voltage-stabilising circuit of the type described above is limited but larger currents can be obtained by use of a current amplifier in conjunction with a voltage-reference diode. One possible circuit is

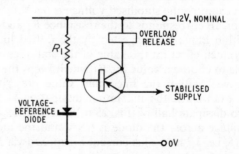

Fig. 16.3. Voltage-stabilising circuit suitable for supplying large currents

illustrated in Fig. 16.3. This can be regarded as an emitter follower the base of which is held at a constant voltage by a junction-diode circuit of the type described above. Such a circuit can provide a current magnification of, say, 50, and, provided a suitable transistor is used, currents of the order of 1 A can be supplied, the input current to the transistor being of the order of 20 mA.

The output voltage from the emitter follower is stabilised, within limits at approximately the breakdown voltage of the voltage-reference diode. Diodes are available with a wide range of breakdown voltages but it is common practice to use diodes rated at low voltages such as 6·8 V and to obtain higher voltages by connecting a number of diodes in series. As explained on page 17 this technique is used to ensure stability of the output voltage against temperature changes: it also has the advantage of increasing the permissible power rating of the voltage-reference source.

There is a danger that the transistor could be damaged if the output of the stabilised supply were short-circuited for this would cause a large current to flow through the transistor causing excessive dissipation in it. The transistor can be protected against such damage by including in the collector circuit a quick-acting overload release device or a resistor which limits the collector current to a safe value.

Circuits based on this principle are employed in units used to supply direct current, from the mains, to transistorised equipment. These units must satisfy two fundamental requirements:

(1) The voltage output from the unit must be substantially un-
    affected by variations in the current drawn: in other words
    the output resistance of the unit must be low. Appreciable
    resistance can cause coupling between transistors in the
    equipment and this can seriously affect operation: it can
    easily cause instability in an i.f. or r.f. amplifier.

(2) The output voltage must be constant and thus independent of,
    for example, variations in mains voltage. As we have seen,
    the gain of transistor amplifiers is dependent on their mean-
    emitter current which is in turn dependent on the supply
    voltage. To achieve constancy of gain, therefore, the output
    of the supply unit must be highly stable.

The output resistance achieved in the elementary circuit of Fig. 16.3 is, in general, not low enough to satisfy requirement (1) and a better performance is achieved by including additional stages of amplification between the voltage-reference diode and the emitter follower. A circuit including three transistors is illustrated in Fig. 16.4 which also illustrates the mains transformer, rectifier and smoothing components.

The diode D1 is biased by R_1 to the breakdown state in which the voltage across it is substantially independent of the current through it. Thus the emitter of TR1 is held at a constant potential. The base of TR1 is connected to the potential divider $R_3R_4R_5$ which applies to the base a certain fraction of the output voltage of the unit. Thus TR1 compares the output voltage with the constant potential from the diode and, if the difference between these voltages varies, amplifies the variations to give a larger variation in TR2 base potential. TR2 is directly coupled to TR3 to form a Darlington circuit used as an emitter follower and TR3 emitter

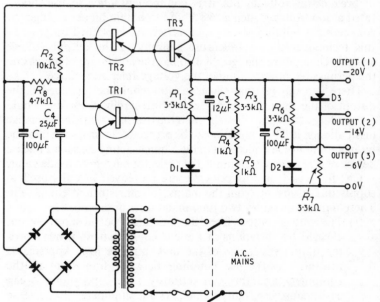

Fig. 16.4. Circuit diagram for a stabilised mains unit suitable for feeding transistorised equipment

potential is the output potential of the unit. Thus the output voltage changes if there is any change in TR1 base-emitter potential: moreover the change in output potential is such as to offset the initial change. For example, if the current taken from the output is suddenly increased, there is a tendency for the output voltage to fall. This reduces the voltage across R_5 and thus the current in TR1, causing its collector potential to go negative. This increases collector current in TR2, thus increasing the output current of the unit: it also increases the output voltage of the unit. This latter change again affects the voltage across R_5: this is, in fact, an example of negative feedback and in the final state of equilibrium the unit supplies the increased current and the fall in output voltage is much smaller than without TR1, TR2 and the feedback loop.

The potential divider $R_3 R_4 R_5$ governs the output voltage of the unit. For example, if D1 has a breakdown voltage of 6·8 V the base potential of TR1, assuming this to be a silicon transistor, is approximately -7.5 V. If an output voltage of 20 V is wanted, the potential divider must have a step-down ratio of approximately $2.7:1$. If the ratio of the potential divider is made adjustable between $2:1$ and $3:1$, the output voltage can be set between 15 V and 22.5 V. The output voltage of the rectifier must, of course, exceed 22.5 V sufficiently to enable TR2 and TR3 to function well. The capacitor C_3 effectively short-circuits the upper arm of the potentiometer at the ripple frequency (100 Hz because full-wave rectification is used) thus increasing feedback at this frequency, so minimising ripple on the d.c. outputs.

We can assess the stability achieved in this circuit in the following way. Suppose the collector current of TR1 is 1 mA: its mutual conductance is then approximately 40 mA/V (see p. 34) and a change of base potential of $1/40$ V, i.e. 25 mV is needed to change the collector current by 1 mA. If we assume the current gain of TR2, TR3 to be 1,000, the change in output current is 1 A for 25 mV change in TR1 base potential. If the potential divider $R_3R_4R_5$ has a step-down ratio of $3:1$, then the change in output voltage is 75 mV. The output resistance of the unit is thus 75 mV/1 A, i.e. 0.075 Ω. Values as low as 0.01 Ω can be achieved by the inclusion of a fourth transistor, Darlington-coupled to TR2 to form a three-stage emitter follower.

For successful operation of this form of stabiliser the current output of TR1 must be transferred without significant loss to TR2 base: thus R_2 must be large compared with the input resistance of the emitter follower TR2, TR3. This is normal technique, of course, in the inter-transistor couplings of a current amplifier. But R_2 has to carry the collector current of TR1 together with the base current of TR2 and if R_2 is large the voltage across it may be several volts. This is also the voltage across TR3 and, with the output current of the unit, determines the dissipation in TR3. If the current to be delivered is small, say 25 mA, the dissipation in TR3 may be well below the permissible maximum and the circuit is satisfactory in the form shown in Fig. 16.4. If, however, the output current is much larger, say 1 A, then the dissipation in TR3 may be excessive in this circuit arrangement. This difficulty can be overcome by keeping the voltage across TR3 very low and by providing a separate stabilised supply purely for R_2: this can be obtained from another winding on the mains transformer, feeding a rectifier, smoothing capacitors and a voltage-reference diode.

Frequently mains units are required to deliver a number of outputs at different stabilised voltages. Often only a small current is required at certain of these voltages and it is possible to use simple stabilising circuits of the type shown in Fig. 16.1 to deliver these small currents.

Two such outputs are (2) and (3) in Fig. 16.4: if the reference-voltage diodes D2 and D3 have breakdown voltages of 6 V and if the large-current output (1) is at -20 V, output (2) is at -14 V and output (3) at -6 V.

The output transistor TR3 must be capable of supplying the maximum current required by the load and is usually mounted on a heat sink to keep its operating temperature at an acceptable value.

CAPACITANCE-DIODE A.F.C. CIRCUIT

As mentioned in Chapter 1 the capacitance of a reverse-biased junction diode varies with the bias voltage. Such a diode can therefore be used for automatic frequency control and Fig. 16.5 gives a circuit diagram which can be used in an f.m. receiver for this purpose. Not all junction diodes are suitable for this application: for some

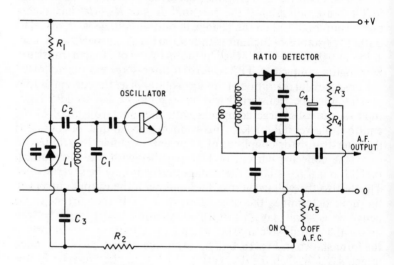

Fig. 16.5. Circuit illustrating the use of a capacitance diode to give a.f.c.

types the damping due to the resistive component of the diode impedance may be sufficient to reduce the oscillation amplitude to a low value or even to prevent oscillation altogether. Diodes with very low damping have been developed for use in a.f.c. circuits.

L_1C_1 is the oscillator tuned circuit and the junction diode is connected across the circuit via the isolating capacitors C_2 and C_3. The diode is reverse-biased from the supply by the potential divider

R_1R_2 and provided that R_1 is reasonably high in value, say, more than 40 kΩ, the damping of the oscillator circuit by this resistor should not seriously reduce the oscillation amplitude. The capacitance of the diode is effectively in parallel with C_1 and alteration in diode bias causes an alteration in oscillator frequency. To obtain a.f.c. the diode bias must be controlled automatically by the degree of mistuning and this can be achieved by returning R_2 to the d.c. output of the discriminator. If the discriminator is a Foster–Seeley type, R_2 may be connected directly to the detector output provided the connection is made on the detector side of the output coupling capacitor. If a ratio detector is used the connection of R_2 is not so straightforward.

The d.c. output of a ratio detector circuit of the type illustrated in Fig. 16.5 has two components: one component varies with tuning and reverses in polarity at the correct tuning point, the relationship between voltage and frequency being substantially linear over a limited frequency range. This is the component responsible for the a.f. output of the detector and is the component required for a.f.c. purposes. The second component is a positive voltage obtained from the long time constant circuit $R_3R_4C_4$: this also varies with tuning but has a maximum at the correct tuning point. To obtain good a.f.c. performance it is advisable to eliminate the second component from the d.c. output of the detector. This can readily be done by earthing the mid-point of the resistor as shown in Fig. 16.5. This gives what might be termed a 'balanced' form of ratio detector which gives zero output voltage at the correct tuning point: the d.c. output of such a detector varies with tuning in the same manner as that of a Foster–Seeley discriminator and R_2 can then be returned to the detector as shown in Fig. 16.5. The junction diode bias must not be affected by a.f. signals in the detector output and these are therefore prevented from reaching the diode by the capacitor C_3 which forms with R_2 a potential divider which considerably attenuates all audio frequency signals.

A.f.c. circuits of this type can be extremely effective, reducing mistuning effects by a factor of as much as 10:1. Manual tuning can be very difficult with a.f.c. and it is desirable to have some means of switching a.f.c. off whilst tuning is being carried out. As soon as the wanted signal is tuned in, a.f.c. is switched on to minimise subsequent tuning drift. Fig. 16.5 indicates one method of switching a.f.c. off. The resistor R_5 is approximately equal to the d.c. resistance of the ratio detector. Such a resistor is necessary to enable the a.f.c. to be switched off and on when the receiver is accurately in tune, without alteration of the bias across the capacitance diode.

USE OF TRANSISTOR TO INCREASE RELAY SENSITIVITY

Electromagnetic relays enable one or more circuits to be switched on and off by a controlling current much smaller than the controlled current. For example a relay requiring 5 mA of input current can control a circuit carrying 5 A.

By use of a transistor as a current amplifier, the sensitivity of a relay can be greatly increased. As we have seen, a common-emitter amplifier can easily give a current gain of 50 and by using such an amplifier with the relay mentioned above only 100 µA of input current is needed to control the 5 A circuit.

A suitable circuit diagram is given in Fig. 16.6. The base of the

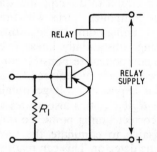

Fig. 16.6. Use of a transistor to increase relay sensitivity

transistor is returned to the emitter via R_1 causing a collector current too small to operate the relay. A small input current causes the transistor to conduct and the relay to operate.

A desirable practical precaution is to connect a diode across the relay winding to prevent generation of large collector voltages across the relay coil when the transistor input is removed and the collector current is cut off. Such voltages can exceed the collector breakdown voltage and can damage the transistor.

The use of a transistor with a relay is a convenient circuit arrangement because the transistor requires no supplies other than that required for the relay and takes up very little space.

D.C. CONVERTERS

Transistors are particularly useful in the construction of d.c. converters, units which can be made remarkably compact and which can convert power from a low voltage source (e.g. 6 V) to a higher voltage (e.g. 120 V) with an efficiency which can approach 85 per cent and is seldom less than 60 per cent.

The transistor in such a converter is used as a switch which interrupts the d.c. supply from the low-voltage source to produce alternating current. This is stepped up in voltage by a transformer or resonant circuit to give a high-voltage supply which is rectified and smoothed to obtain the high-voltage output. For high efficiency the power dissipated in the transistor itself must be small. The power dissipated in a transistor is of course given by the product of the collector current and the collector-emitter voltage. The power is therefore small when the collector current is nearly zero, that is to say when the transistor is cut off and also when the collector-emitter voltage is nearly zero, that is to say when the transistor is fully conducting. Thus the design of the d.c. convertor must be such that the transistor is always either fully conducting or cut off. This is achieved by using the transistor as an astable relaxation oscillator which generates rectangular waves.

The circuit of one type of d.c. converter is given in Fig. 16.7. The

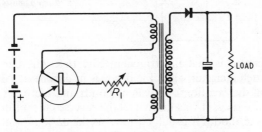

Fig. 16.7. Complete circuit of transistor d.c. converter

collector and base circuits of the transistor are coupled to give positive feedback and consequent oscillation. Considerable feedback is necessary to drive the transistor hard into conduction and cut off. The ratio of the periods of conduction and non-conduction can be controlled by adjustment of the value of the resistor R_1, which determines the base bias of the transistor. Frequently d.c. converters employ two transistors operating in push-pull.

Oscillation frequencies in d.c. converters may lie between 500 Hz and 10 kHz. If a converter is required to be particularly compact the transformer and smoothing capacitor must be small. This is practicable, provided the working frequency is high, and the tendency is therefore to have high working frequencies in compact converters.

PHOTO-DIODE

It is shown in Chapter 1 that the current which flows across a reverse-biased pn junction is carried by minority carriers, i.e. by the electrons and holes liberated by breakdown of the covalent bonds of the intrinsic semiconducting material. This current is substantially independent of the reverse bias voltage, provided this exceeds approximately 1 V, but can be increased by heating the material or by allowing light to fall on it: both give the semiconductor atoms more energy and cause more covalent bonds to break. Where sensitivity to light is undesirable junction diodes and bipolar transistors are sealed in opaque containers: if sensitivity to light is required a transparent container is employed.

A junction diode in a transparent container is known as a photo-diode and can be used to indicate the presence of light. An obvious

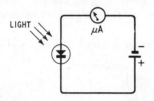

Fig. 16.8. Simple light-meter circuit
using a photo-diode

form of circuit is that illustrated in Fig. 16.8. The current which flows in such a circuit when the diode is in darkness is due entirely to thermal dissociation of covalent bonds and increases rapidly as temperature rises. It was known as the reverse current in Chapter 1 but in photo-diodes is usually known as the dark current. The ratio of light to dark current thus decreases as temperature rises.

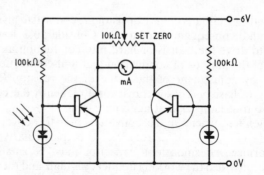

Fig. 16.9 Light-meter circuit employing two photo-
diodes and two transistors

The output power from a photo-diode is limited and amplification is essential if greater power is required, e.g. to operate a milliammeter or a relay. Amplification can be provided by a transistor direct-coupled to the photo-diode as shown in Fig. 16.9. In this circuit two photo-diodes and two transistors are used in a balanced circuit which largely eliminates the effects of temperature changes and gives a meter reading which depends only on the illumination falling on one of the photo-diodes.

To set up the circuit the two photo-diodes are screened from light and the potentiometer is adjusted to give zero meter reading. When one of the photo-diodes is now exposed to light the meter gives an indication proportional to the illumination and the meter can, in fact, be calibrated in terms of illumination.

PHOTO-TRANSISTORS

A photo-transistor may alternatively be used to produce an output greater than is possible from a photo-diode. The mechanism of the amplification inherent in a photo-transistor can be explained in the following way.

First consider a transistor connected to a supply as indicated in Fig. 16.10(a). It was pointed out at the beginning of Chapter 6 that the small leakage current I_{CBO} which flows in such a circuit arises from

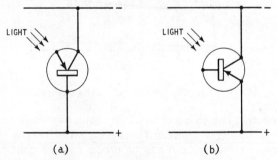

Fig. 16.10. Photo-sensitivity of a transistor arranged as at (b) is considerably greater than when arranged as at (a)

dissociation of covalent bonds and increases rapidly as temperature rises. This current also increases if light falls on the transistor because this also breaks up covalent bonds. The leakage current in this circuit is that of the reverse-biased collector-base junction and is of the same order as that of a photo-diode. Such a transistor circuit therefore provides no greater output than is available from a photo-diode.

Now consider the transistor connected to the supply as shown in Fig. 16.10(b). Chapter 6 shows that the leakage current I_{CBO} for such a circuit is $(\beta + 1)$ times I_{CBO}, i.e. is much greater than that of the transistor connected as in Fig. 16.10(a). It is in fact this leakage current which necessitates protective circuits for d.c. stabilisation in common-emitter amplifiers using germanium transistors. The leakage current may be due to hole-electron pairs released by heat or released by light and thus a transistor used in a circuit arrangement such as that of Fig. 16.10(b) can produce a considerable increase in collector current when light falls on the base region. This is illustrated in Fig. 16.11 which gives the collector current-collector voltage characteristics for a photo-transistor plotted with incident light as

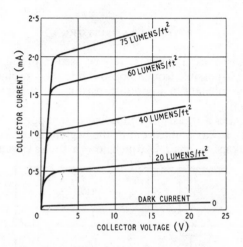

Fig. 16.11. Characteristics of a photo-transistor

the parameter. The characteristics are similar in shape to those of a common-emitter amplifier. The curves show that a change of collector current of 0·5 mA can be produced by a change in light input of 20 lumens/ft^2.

If the base circuit of the photo-transistor is open-circuited as shown in Fig. 16.10(b), the variations in collector current due to temperature changes are considerable. Thus the ratio of light current to dark current is limited and may vary in practice from 100 at 25°C to 10 at 45°C. Provided the photo-transistor is not required to work in surroundings where wide variations in temperature are likely to occur the very simple circuit of Fig. 16.10(b) may be satisfactory.

Where wide variations in temperature are inevitable it is preferable to have a larger ratio of light current to dark current. This can be achieved by use of a resistor connected between base and emitter as shown in Fig. 16.12. With a resistor of approximately 5 kΩ the ratio of light current to dark current is now increased to 400 at 25°C falling to 20 at 45°C.

There are some applications of photo-electric devices where the light input is 'chopped'; this occurs, for example, where the device is used for counting articles moving along a conveyor belt. A photo-transistor circuit suitable for such an application is illustrated in

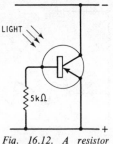

Fig. 16.12. A resistor connected between base and emitter of a photo-transistor can be used to improve the ratio of light current to dark current

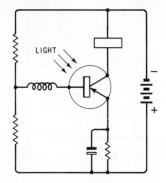

Fig. 16.13. Circuit using a photo-transistor with an inter-rupted light input

Fig. 16.13. This employs the potential-divider method (see Chapter 6) of dark-current stabilisation but to avoid reduction in light current an inductor is included in series with the base lead to the photo-transistor. To give maximum output this inductor should be parallel-resonant at the frequency of the light variation.

UNIJUNCTION TRANSISTOR (DOUBLE-BASE DIODE)

This device has a filament of, say, n-type silicon with ohmic contacts at each end and a p-type junction near the centre. If the junction is reverse-biased the filament may have a resistance of, say, 10 kΩ but this can be substantially reduced by forward biasing the junction. The device makes possible the simple pulse-generating circuit shown in Fig. 16.14.

For simplicity we will assume that the base contacts are con-nected to a 10-V source and that the pn junction is at the centre of the filament. Thus the base at the junction has a bias of -5 V. Initially the capacitor C is uncharged and the emitter potential is -10 V. The pn junction is thus reverse-biased and a steady current of 1 mA flows through the base via R_1 and R_2 (both assumed small compared with 10 kΩ).

As *C* charges through *R* the potential of the emitter moves positively, following an exponential law, until it reaches the value −5 V. At this moment the pn junction has zero bias and any further rise in voltage across *C* causes the junction to conduct. The charge carriers are chiefly holes which are injected by the emitter p-region across the junction into the base n-region. Here the holes tend to

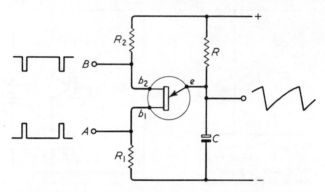

Fig. 16.14. Simple pulse-generating circuit using a unijunction transistor, i.e. double-base diode

move towards the most negative part of the base region and in doing so reduce the effective resistance of the lower half of the base region. This in turn increases the current crossing the junction and encourages the injection of further holes. Thus a regenerative process is set up which culminates in the rapid discharge of *C* by the current flowing across the junction and through the lower half of the base and R_1. The burst of current in R_1 generates a positive-going pulse at terminal *A*. The momentary reduction in base resistance due to hole injection causes a transient increase in current in R_2 and thus generates a negative-going pulse at terminal *B*.

The collapse in voltage across *C* causes the pn junction to become reverse-biased again and the circuit is back in the original state. The cycle then restarts and continues at a rate dependent on the time constant RC. This simple circuit thus gives pulses of both polarities and an approximation to a sawtooth output.

THYRISTOR (CONTROLLED SEMICONDUCTOR RECTIFIER)

This is a four-layer semiconductor device with a structure of the form shown in Fig. 16.15, which can be represented more simply as in Fig. 16.16. It has two stable states, one in which the resistance is very low (the conductive state) and the other in which the resistance is very high (the non-conductive state). The device can be switched very rapidly from non-conduction to conduction and very little power is needed to bring about this change of state. Thus the

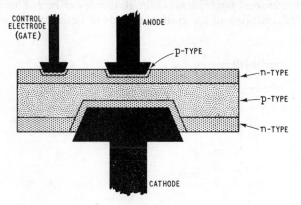

Fig. 16.15. *Construction of a controlled semiconductor rectifier*

pnpn transistor has properties similar to those of a thyratron (gas-filled valve) but it is far more efficient; the device is used mainly for switching and power control purposes, e.g. as a controlled rectifier.

To understand the mode of action of the pnpn transistor, the device can be regarded as made up of two bipolar transistors, one of pnp type and the other of npn type, direct-coupled as shown in

Fig. 16.16. *Reverse-biased
controlled rectifier*

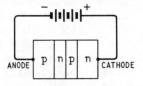

Fig. 16.17. Suppose a voltage is applied to the outermost regions of the device, as shown in Fig. 16.16. The polarity of this voltage is such as to reverse bias the two outer junctions of the device: these are the base-emitter junctions of the two bipolar transistors and both are thus cut off. Very little current can flow through the device and with this polarity of applied voltage the pnpn transistor is in its non-conductive state.

Now suppose the polarity of the applied voltage is reversed: the positive terminal of the supply is now connected to the outermost p-region (anode) and the negative terminal to the cathode. The applied voltage now biases the two outer junctions in the forward direction and the outermost p- and n-regions can now act as emitters in the two bipolar transistors. The centre junction (which is the base-collector junction for both transistors) is reverse-biased and most of the applied voltage acts across this junction.

Let the current which flows through the device be I. This is then the current flowing in the emitter regions of both the pnp and the

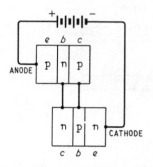

Fig. 16.17. Forward-biased rectifier represented as two direct-coupled transistors

npn transistors (see Fig. 16.17). The current crossing the centre junction of the controlled rectifier is, however, made up of the following three components:

(1) The current crossing the base-collector junction of the pnp transistor. This component is carried by holes which originate in the emitter region, cross the base region and enter the collector region. The fraction of the total number of holes emitted which reach the collector region is α_1, the current amplification factor of the pnp transistor. The current reaching the collector region is $\alpha_1 I$.

(2) The current crossing the base-collector junction of the npn transistor. The component is carried by electrons which originate in the emitter region, cross the base region and enter the collector region. The fraction of the total number of electrons emitted which reach the collector region is α_2, the current amplification factor of the npn transistor. The current reaching the collector region is $\alpha_2 I$.

(3) The leakage current of both transistors, assumed to total I_{CBO}. Thus we have

$$I = \alpha_1 I + \alpha_2 I + I_{CBO}$$

from which

$$I = \frac{I_{CBO}}{1 - (\alpha_1 + \alpha_2)}$$

This shows that provided $(\alpha_1 + \alpha_2)$ is less than unity, I has a finite value. For example, if $\alpha_1 + \alpha_2 = 0.9$, $I = 10 I_{CBO}$ which is a very small current. For such values of current amplification factor the device is stable though forward-biased.

If, however, the alphas total unity (each equal to 0.5, say) the transistor is on the verge of instability. Any current amplification occurring in the reverse-biased centre junction (due to ionisation by collision, for example) increases both α_1 and α_2, causing their sum to exceed unity. Equilibrium is then impossible and the current I rises rapidly to a very high value. The process is regenerative because any increase in I causes enhanced ionisation, which in turn increases I. Such breakdown occurs naturally if the applied voltage is raised to a high value (causing a sufficiently intense electric field across the centre junction) and in a controlled rectifier the design might be such that breakdown occurs at an applied voltage of 350 V. The device is now in its conductive state.

So far we have assumed that there is no external connection to the inner p- or n-region. Suppose that such a connection is used to apply a positive voltage to the inner p-region (relative to the cathode) and that this supplies a current of I_b electrons to the device. It is a property of transistors, particularly those employing silicon, that the current gain α depends on the emitter current, being small for small emitter currents and increasing as the emitter current is increased. The application of a forward bias current to the inner p-region increases the emitter current from the outer n-region, so increasing the current gain α_2 of the npn transistor and the sum of the two alphas. Breakdown now occurs more easily, i.e. at a lower applied voltage than in the absence of external bias. Moreover, by increasing I_b, the breakdown voltage can be reduced as low as desired: in practice, the breakdown voltage can be reduced to a few volts only.

The characteristics of the controlled rectifier are given in Fig. 16.18: they are similar to those of a gas-filled triode valve (thyratron). The greatest control power required is only 5 V at 100 mA: this can control a current of 50 A at 250 V. With no control signal the rectifier presents a very high resistance and an applied voltage of 350 V is required to cause breakdown. The resistance then falls to a very low value and currents of up to 50 A can flow through the rectifier with less than 2 V drop across it. Once breakdown has occurred, the control signal may be removed but the low resistance will remain until the forward current in the rectifier has fallen to a low value. Thus the device can be fired by very short-duration pulses applied to the control terminal. When firing occurs the build-up of current in the rectifier can be very rapid: rise times of 1 μs can be achieved.

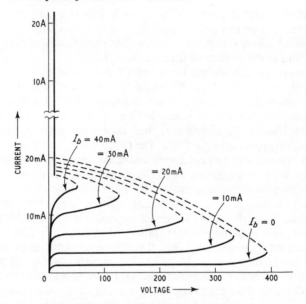

Fig. 16.18. Anode characteristics of a controlled rectifier

The controlled rectifier can be used in a.c. power control circuits, in d.c. converters and in voltage-regulated d.c. supply circuits.

The general symbol for a controlled semiconductor rectifier is

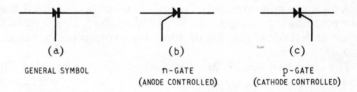

Fig. 16.19. Graphical symbols for controlled semiconductor rectifiers

given in Fig. 16.19(a). A rectifier with an n-gate, i.e. one designed for a negative-going triggering signal applied to the inner n-region can be represented by the symbol of Fig. 16.19(b): such a rectifier is sometimes called anode-controlled. A rectifier with a p-gate, i.e. one designed for a positive-going triggering signal applied to the inner p-region can be represented by the symbol of Fig. 16.19(c): such a rectifier is sometimes termed cathode-controlled.

Inverter Using Controlled Rectifiers

Controlled rectifiers can be used in inverters, i.e. equipments which generate alternating supplies from a direct-current supply. Inverters enable mains-voltage apparatus, e.g. fluorescent lamps, to be operated where only a low-voltage d.c. supply is available.

The circuit diagram for one type of inverter is given in Fig. 16.20.

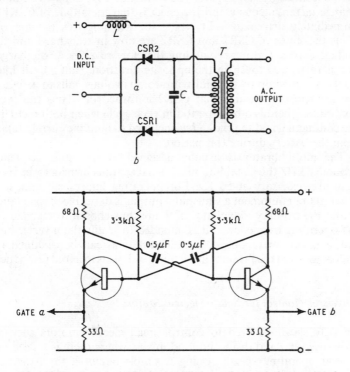

Fig. 16.20. An inverter using controlled rectifiers and the associated multivibrator switching circuit. Component values shown are suitable for a frequency of around 1 kHz

This has two controlled rectifiers which conduct alternately to interrupt a d.c. supply at a rate determined by an astable multivibrator. The interrupted supply is fed to the primary winding of a step-up transformer *T*, from the secondary of which the high-voltage a.c. supply is obtained.

The multivibrator is of the collector-coupled type illustrated in Fig. 13.8 and the outputs, taken from resistors in the emitter circuits of TR1 and TR2, are directly applied to the gates of the controlled rectifiers. Consider conditions in the circuit during the period when CSR1 is on. The voltage across CSR1 is negligibly low and the full battery voltage V appears across the lower half of T primary winding. By auto-transformer action an equal voltage appears across the upper half of this winding and thus the capacitor C is charged to $2V$. When CSR2 is abruptly switched on by the multivibrator, the voltage across it suddenly falls to zero. The voltage across a capacitor cannot change instantaneously and the effect is that the anode of CSR1 is immediately driven to $-2V$ volts, so switching this rectifier off. Thus the states of CSR1 and CSR2 are now interchanged and the voltage across T primary winding therefore reverses. C discharges and is recharged to $2V$ in the opposite direction. Half a cycle later CSR1 is turned on by the multivibrator, its voltage falls to zero and the capacitor C ensures that CSR2 is turned off. Thus the cycle continues. There is a brief period in each cycle when both rectifiers are conducting and the choke L is included to limit the current taken from the battery during this period.

The multivibrator is usually designed to run at a frequency around 1 kHz thus enabling much smaller transformers to be used than are necessary at 50 Hz. For operating fluorescent lamps or other mains equipment a sinusoidal output is desirable: capacitance C and the primary inductance of T are thus chosen to resonate at the operating frequency. If it is intended to rectify the inverter output, a square-wave output is preferable and can be obtained by choosing C and the primary inductance of T so as to avoid resonance.

Thyristor Control for Small Electric Motors

Thyristors can be used to control small electric motors such as those used in hand drills and food mixers where there is a need for several operating speeds. Using a simple circuit a thyristor can determine the motor speed by control of the fraction of each cycle during which mains current is allowed to flow (known as the conduction angle) and the speed can be maintained, in spite of varying mains voltage or changing mechanical load on the motor, by feedback from the back e.m.f. from the motor.

The basic circuit is shown in Fig. 16.21. When the thyristor anode is negative with respect to the cathode, the thyristor is non-conductive: thus for one-half of each mains cycle the motor receives no power. Nevertheless the motor is rotating as a result of the power received during the other half-cycle and generates a back e.m.f. proportional to speed (and residual flux) which biases the thyristor

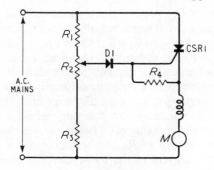

*Fig. 16.21. Basic circuit diagram for control
of electric-motor speed by a thyristor*

cathode positively. When the thyristor anode is positive with respect
to the cathode, the thyristor conducts if the gate is suitably biased:
thus for this half of each mains cycle the motor receives power
controlled by the gate circuit. The diode D1 conducts positive-going
trigger voltages to the gate but when the thyristor conducts, the
thyristor cathode potential takes up the anode value so reverse-
biasing D1 and isolating the thyristor from the trigger circuit. The
gate voltage must exceed the cathode bias by a fixed amount to turn
the thyristor on and the gate voltage is determined by the setting
of R_2. The more positive the gate is made, the larger is the conduction
angle and the higher the motor speed.

This simple circuit has disadvantages: the chief are that the range
of conduction angle achievable by adjustment of R_2 is limited and
that dissipation in the potential divider is high.

A better circuit is illustrated in Fig. 16.22. The gate circuit operates

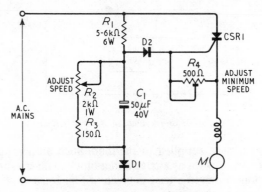

*Fig. 16.22. Practical circuit diagram for control of
electric-motor speed by a thyristor*

as a diode detector (see page 201). On positive half-cycles D1 conducts and charges C_1. On negative half-cycles D1 is non-conductive and C_1 discharges through $(R_2 + R_3)$. On the next positive half-cycle D1 conducts again and the charge lost from C_1 is restored. The extent of the discharge of C_1 depends on the value of $(R_2 + R_3)$ and can be controlled over a wide range by adjustment of R_2. The smaller R_2 is made, the more complete is the discharge of C_1 during negative half-cycles and the longer is the charging period during positive half-cycles. In fact the charging of C_1 begins early in the positive half-cycle and by adjustment of R_2 can be made to last from a small fraction to three-quarters of the half-cycle. The charging current flows through R_1 and generates across it a negative-going pulse which reverse-biases D2, isolating the thyristor gate and preventing the thyristor from conducting. Thus the thyristor becomes conductive as the charging current ceases and is cut off when the anode voltage swings to zero at the end of the positive half-cycle. R_2 controls the duration of the charging process and hence the conduction angle of the thyristor.

The Triac

For controlling larger motors and other large a.c. loads, the circuit of Fig. 16.22 is further improved by using a second thyristor arranged to conduct when the first thyristor is non-conductive. In this way

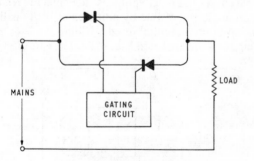

Fig. 16.23. Block diagram of a triac circuit

more power can be supplied to the load. Such an arrangement of two thyristors is known as a triac and a block diagram showing the essential connections to a triac is given in Fig. 16.23.

THE F.E.T. AS A VARIABLE RESISTOR

The internal drain-source path in an f.e.t. differs in nature from the corresponding path in a bipolar transistor or a valve. It has no rectifying properties but consists simply of a conducting channel. The drain-source path is, in fact, a linear resistor the value of which can be varied from a low value almost to infinity by adjustment of the gate potential. One application of this, namely the use of the f.e.t. as a switch in digital circuits, is mentioned in Chapter 15.

The fact that the drain-source path is a linear resistor means that alternating voltages such as a.f. and r.f. signals can be applied to it (provided the amplitude is small compared with 1 V). Thus an f.e.t. can be used to construct a voltage-controlled attenuator as suggested in Fig. 16.24. The j.u.g.f.e.t. forms with R_1 a potential divider, the

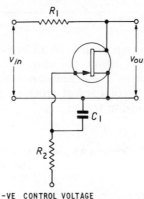

Fig. 16.24. An f.e.t. used as a variable resistance in a voltage-controlled attenuator

attenuation of which can be controlled by adjustment of the negative voltage applied to the gate. R_2C_1 are decoupling components. Because of the very high input resistance of the j.u.g.f.e.t. no power is needed to adjust this attenuator. This principle is useful in constructing remotely controlled attenuators and has obvious applications in a.g.c. systems.

The Manufacture of Transistors
and Integrated Circuits

The two principal raw materials used in the manufacture of transistors are germanium and silicon. We shall first describe the methods used to purify these semiconducting materials and will then give an account of the most important methods of manufacturing germanium and silicon transistors.

PREPARATION OF GERMANIUM FOR TRANSISTOR MANUFACTURE

Germanium is obtained as a byproduct of metal refining, e.g. zinc-refining, in the U.S.A. and can also be obtained from the flue dust of certain coals in the U.K. After extraction by chemical methods, the germanium is usually marketed in the form of the dioxide GeO_2, a white powder.

The first step in the production of germanium for transistor manufacture is the reduction of the oxide to the element. This is achieved by heating the oxide to 650°C in a stream of pure hydrogen. The germanium powder so obtained is melted in an inert atmosphere and then cast into bars.

The bars are not pure enough for use in transistors: moreover, their crystalline form is unlikely to be suitable. Further purification is therefore carried out by the process known as *zone refining*. This relies for its success on the fact that, at the melting point of germanium, most of the impurities are more soluble in the liquid than in the solid form of the element. Thus, if a short length of a germanium bar is melted and if the molten portion is caused to move along the bar, say, from right to left, the impurities tend to follow the movement and concentrate at the left-hand end of the bar. The localised heating is usually carried out by r.f. induction, the germanium bar being supported in an inert atmosphere in a graphite boat which is slowly moved along a silica tube contained within the loops of the r.f. supply. By repeating the zone-refining process a number of times, the purity of the bar (except for the left-hand end, of course) can be raised to the degree required in transistor manufacture.

The single crystals required can be grown by dipping a small seed crystal into a bath of molten zone-refined germanium and slowly withdrawing, i.e. pulling the seed as the crystal grows and cools. Controlled amounts of p or n impurities can be added to the molten material to give the type of germanium required for transistor manufacture.

PREPARATION OF SILICON FOR TRANSISTOR MANUFACTURE

Silicon forms approximately 25 per cent of the earth's crust but it is difficult to extract the element in a form pure enough for use in transistors, primarily because silicon melts at a very high temperature (1,400°C) and it is very reactive when molten, attacking most crucible materials. In particular, the removal of the last traces of boron is most troublesome.

Silicon occurs widely as the dioxide (sand) and the first stage in the extraction is the reduction of this oxide in an arc between carbon electrodes. The low-grade silicon so obtained is purified by a number of chemical processes and is then subjected to a zone-refining process similar to that used for germanium. To avoid contamination from containing vessels, however, the silicon, in the form of a bar, is supported vertically by its ends whilst heated locally by r.f. induction—a technique known as *vertical* or *floating zone refining*. Finally crystals of silicon can be obtained by the method of pulling or growing.

GROWN TRANSISTORS

The first method of manufacturing transistors was the so-called *dope-growing* method described by Shockley in 1951. In this method the molten semiconductor is first treated with sufficient n-type impurity to give the required collector resistivity (say, 1 to 2 Ω-cm) and an n-type ingot of the required crystalline form and resistivity is obtained by pulling or growing as described above. After a suitable length of crystal has been grown, a pellet of p-type impurity is added to the molten material and this is of sufficient mass to neutralise the n-type impurity and to give the p-type resistivity required in the base region (say, 1 to 2 Ω-cm). The ingot is now grown only a very short distance (equal to the thickness of base layer required— say, 0·001 in) and a second pellet, this time of n-type impurity is added to the molten material to give the required resistivity of emitter region—say, 0·01 Ω-cm or less. After a suitable length of the emitter region has been grown the ingot is withdrawn from the molten material.

The ingot (Fig. A.1) is now cut in directions parallel to the

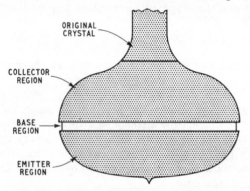

ORIGINAL CRYSTAL

COLLECTOR REGION

BASE REGION

EMITTER REGION

Fig. A.1. A grown crystal

direction of growth to produce transistors approximately 0·02-in square and 0·2-in long. Several hundred transistors can be obtained from one ingot. The emitter and collector ends can be distinguished by etching the transistors with a material which affects these ends at different rates. Connecting leads can then be soldered to the emitter and collector regions but the connection to the thin base region is difficult to make because, of course, this region cannot be seen. It can, however, be detected by exploring the surface of the transistor with a fine wire contact which is included in a circuit designed to give a signal when the base region is touched. An electric discharge can then be used to secure the wire to the base region.

ALLOY JUNCTION TRANSISTORS

The starting point of this method of manufacture of germanium transistors is a wafer of crystalline semiconductor, say, n-type, of

5 Ω-cm resistivity. The wafers are approximately 0·1-in square and 0·003-in thick. A pellet of a group-III element such as indium is placed at the centre of each face of the wafer which is then heated to a few hundred degrees Centigrade, well above the melting point of indium (155°C) but well below that of germanium (940°C). The indium melts and continues to dissolve germanium until a saturated solution is obtained. The wafer is now allowed to cool slowly and the dissolved germanium crystallises out. The recrystallised germanium retains some indium and thus forms a region of p-type germanium on either face of the wafer as shown in Fig. A.2. These

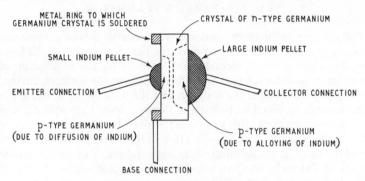

METAL RING TO WHICH
GERMANIUM CRYSTAL IS SOLDERED

CRYSTAL OF n-TYPE GERMANIUM

SMALL INDIUM PELLET

LARGE INDIUM PELLET

EMITTER CONNECTION

COLLECTOR CONNECTION

p-TYPE GERMANIUM
(DUE TO DIFFUSION OF INDIUM)

p-TYPE GERMANIUM
(DUE TO ALLOYING OF INDIUM)

BASE CONNECTION

Fig. A.2. Construction of a germanium alloy-junction pnp transistor

regions are separated by a region with the n-type conductivity of the original material, thus giving a pnp transistor. The final diameters of the indium regions are commonly of the order of 0·015 in (emitter) and 0·03 in (collector). Connections are soldered to both regions and the base connection is often made to a metal ring soldered to the wafer and closely surrounding the emitter region. After cleaning the assembly is hermetically sealed in a light-proof container.

A similar technique can be used to manufacture npn germanium transistors: a wafer of p-type germanium is used and pellets of a group-V element such as arsenic or antimony, carried in a neutral (group-IV) element such as lead, are alloyed to it.

Very large numbers of pnp germanium transistors have been manufactured by this method and they can operate up to frequencies of the order of 5 to 7 MHz but to obtain reasonably consistent base thicknesses close control is required over the thickness of the wafers and the time and temperature of the processing.

Silicon transistors can also be made by this method, the usual alloying metal being aluminium but the very different coefficients of expansion of aluminium and silicon have caused difficulties.

The principal factor limiting the performance of a transistor at high frequencies is the time taken for charge carriers to cross the base region and to obtain a good performance this time must be reduced to a minimum. The obvious way to improve the high-frequency performance is to reduce the thickness of the base region even though this may limit the maximum permitted collector voltage to a few volts. This is the technique employed in the surface-barrier transistor pioneered by Philco in 1953.

SURFACE-BARRIER TRANSISTOR

The starting point of the process is a wafer of n-type germanium to which a ring is soldered to form the base connection. Both faces of the wafer are then etched electrolytically by jets of solution of precise cross section which are directed against the surfaces, current being passed between the solution and the wafer so as to cause the germanium to dissolve. Etching is stopped when the required base thickness is reached: this can be determined by the transparency of the base region to light. The direction of the current is now reversed and the metal in the solution (commonly indium) is deposited on the faces of the wafer to form emitter and collector regions to which connections are made. This process is well suited to automatic operation and large numbers of transistors have been produced. They are not so robust as those produced by other processes but operate at frequencies up to 50 MHz. The robustness of the transistors can be improved by lightly alloying the layers of indium to the base region by heating: the resulting transistors are termed *micro-alloy* types.

Another method of reducing transit time in a transistor is to vary the impurity concentration in the base region so as to produce an electric field which aids the passage of charge carriers across the region. To give this effect the impurity concentration must be a maximum near the emitter junction and a minimum near the collector junction; an exponential distribution or one approaching exponential form is desirable.

The technique of solid-state diffusion may be employed to produce such a graded base region. One method is to expose the semiconductor material to a vapour of the desired impurity in a furnace. This causes impurity atoms to diffuse into the crystal structure to give an impurity concentration which falls off as depth of penetration increases. If a germanium wafer with such a graded impurity concentration is used as the starting point in the manufacture of a micro-alloy transistor, a much-reduced transit time can be obtained and the transistor has a better high-frequency response than one with a uniform base region. In theory the cut-off frequency of a transistor can be improved by up to eight times by grading the impurity concentration in the base region but in practice the improvement is normally less than this. Graded-base micro-alloy transistors, known as *micro-alloy diffused transistors* (MADTs) operate at frequencies up to 300 MHz.

DRIFT TRANSISTORS

Wafers of graded impurity concentration can also be used as the starting point in making alloy junction transistors. The resulting transistors are known as drift types and operate at frequencies up to 100 MHz.

DIFFUSED TRANSISTORS

The technique of solid-state diffusion was mentioned above as a means of producing a semiconductor with a graded impurity concentration. This technique is also extensively used as a means of producing layers of p-type and n-type conductivity of very thin but controllable thickness which can be used as the emitter, base and sometimes collector regions of a transistor. There are many methods of manufacturing transistors using such layers: two will now be described.

In one method the starting point is a wafer of p-type germanium which ultimately forms the collector region. To this, two pellets are alloyed in close proximity: one pellet is of a group-III element such as antimony and the other is of mixed group-III and group-V elements (e.g. antimony and aluminium). The wafer is now heated

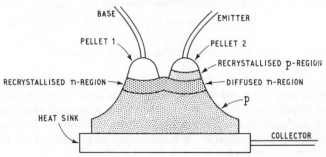

Fig. A.3. Construction of PADT pnp germanium transistor

causing the elements in the pellets to diffuse into the germanium. Only n-type impurities diffuse from the first pellet but both n-type and p-type from the second. However, in p-type germanium, n-type impurities diffuse more rapidly than the p-type and thus, after diffusion, we obtain a structure such as that illustrated in Fig. A.3: the two diffused n-regions merge to form a base layer (to which the first pellet gives ohmic contact) whilst the recrystallised p-region acts as the emitter region (to which the second pellet gives ohmic contact). Finally the diffused n-type layer is etched away except around the two pellets which are masked during this operation. This leaves the active part of the transistor in the form of a mesa projecting from the collector material as shown in the diagram. This particular method of manufacture yields transistors known as *post-alloy diffused* types (PADTs).

The difficulty of manufacturing alloy-junction silicon transistors has already been mentioned. Many silicon transistors are now manufactured by other techniques which include solid-state diffusion. In one method the starting point is a crystal of n-type silicon with a uniform resistivity of, say, 1 Ω-cm. This is cut into slices about 0·015-in thick each of which finally yields 100 or more transistors. The slices are heated to around 1,250°C in a furnace with a source of p-type impurity. The impurity volatilises to produce an atmosphere which diffuses into the surface of the slice to a small but controllable depth to give a p-region which is later used as the base areas of the

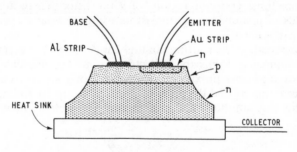

Fig. A.4. Construction of a diffused silicon transistor

transistors. The slice is now covered with a mask containing 100 or more apertures and is again heated, this time in an atmosphere of an n-type impurity. This diffuses to form a number of n-regions which are used as the emitter areas of the transistors. The slice is afterwards cut to separate the individual transistors. A sectional view of one transistor is given in Fig. A.4.

Connections are now made to the three regions of each transistor.

The body of the semiconductor is soldered to a heat sink which provides electrical connection to the collector. Contact to the emitter region is obtained by depositing a thin gold strip on the region by evaporation via a mask and finally lightly alloying it to the n-region by heating. The base connection is obtained by similarly depositing a thin strip of aluminium on the exposed p-region and once again lightly alloying it to provide an ohmic contact.

Only the base region between the emitter and collector junctions is vital to the action of the transistor and surplus base-collector junction area causes unnecessary collector capacitance. This capacitance is minimised by etching away the base material except that near the gold and aluminium, leaving the active part of the transistor standing up from the collector region in the form of a mesa. Such transistors operate at frequencies up to 500 MHz.

EPITAXIAL DIFFUSED TRANSISTORS

The performance of silicon and germanium diffused transistors manufactured by the above processes is partly limited by the resistance of that part of the wafer forming the collector region. Only the upper crust of the wafer, containing the junctions, is effective in providing transistor action and the thickness of the collector region should be as thin as possible to minimise its resistance. However, it is difficult to work with a thickness less than about 0·003 in, yet such a thickness gives too high a resistance for some applications. If most of the silicon or germanium below the junctions could be replaced by low-resistivity material, the performance of the transistor could be greatly improved. This is effectively achieved by using the technique known as *epitaxy*.

In a silicon epitaxial transistor the starting point is a wafer of very low resistivity, e.g. 0·002 Ω-cm. A thin skin of silicon of resistivity suitable for transistor action is deposited on the face of the wafer in such a way that it takes up the same crystalline structure as the wafer. The added layer is termed the epitaxial layer and is produced by heating the wafer to about 1,200°C in a furnace whilst it is exposed to an atmosphere of, for example, silicon tetrachloride and hydrogen. The hydrogen reduces the chloride to silicon which is deposited on the wafer. At the temperature used, the silicon atoms are mobile and take up their correct orientation relative to the crystalline structure of the wafer. A halide of a group-V impurity is also added to give the epitaxial layer the required n-type conductivity of about 1 Ω-cm. A typical thickness of an epitaxial layer is 10 microns.

The wafer, thus treated, can now be used for the manufacture of diffused transistors by the method described above, the diffusion occurring within the epitaxial layer. The characteristics of the transistors so produced are much better than those of a transistor without the epitaxial layer because of the much-reduced collector series resistance.

A similar technique is also possible with germanium transistors.

PLANAR TRANSISTORS

When a silicon wafer is heated to about 1,200°C in an atmosphere of water vapour or oxygen a skin of silicon dioxide SiO_2 forms on the surface. This skin is a most effective seal against the ingress of moisture at room temperatures and has made possible the method of manufacture of planar transistors which is described below.

A crystal of n-type silicon, about 1 in in diameter, is cut into slices about 0·008-in thick. The slices are lapped and etched to approximately 0·003-in thickness and, if required, an epitaxial layer can be formed on one surface. The slices are now heated in an oxidising atmosphere to acquire a protective coating of silicon dioxide. At this stage each slice has a sectional view similar to that shown in Fig. A.5(a). Each slice yields ultimately up to 1,000 transistors and the next stage is to mark off the individual transistors. This is achieved by a photo-lithographic process: each slice is coated in a dark room with a photo-sensitive material (known as photo-resist) and is then exposed to ultra-violet light via a mask containing an array of apertures corresponding to the base areas of the 1,000 transistors. The slice is now developed to remove the photo-resist from these regions thus exposing the silicon dioxide coating. Next the slice is treated with an etch which removes the silicon dioxide from the exposed regions. The remainder of the photo-resist is now dissolved: the cross-section of the slice now appears as in Fig. A.5(b) which shows a gap in the layer of silicon dioxide defining the base area for a single transistor.

The slice is now exposed at a high temperature to a boron-rich atmosphere. The silicon dioxide coating protects the slice against diffusion of boron except at the exposed areas and here boron diffuses isotropically, i.e. horizontally under the protective coating as well as vertically into the crystal, thus forming a p-type base region. The slice is now returned to the oxidising atmosphere and a coating of silicon dioxide is formed over the base areas (and the rest of the slice) to give a cross-section similar to that shown in Fig. A.5(c).

The emitter areas are now defined by a similar process of masking,

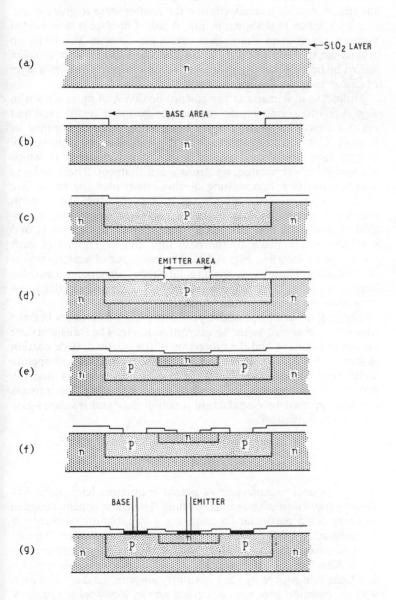

Fig. A.5. Stages in the manufacture of planar transistors

photo-lithography, exposure to ultra-violet light, etching, etc., and the silicon dioxide is removed from the emitter areas to give a cross-section such as that shown in Fig. A.5(d). The slice is now heated whilst exposed to an atmosphere rich in phosphorus. This forms an n-type emitter region by diffusion and the exposed area is again sealed by heating the slice in an oxidising atmosphere to form a layer of silicon dioxide. See Fig. A.5(e).

Holes are now made in the silicon dioxide coating as shown in Fig. A.5(f) to permit ohmic contacts to be made to the base and emitter areas, the position of the holes being again determined by a mask. Contacts are then made to the transistors by a process of evaporation: the slice is placed in a vacuum chamber in which aluminium is evaporated, e.g. from a hot filament. This results in a deposition of a thin coating of aluminium over the entire face of the slice. Finally the aluminium is removed from the areas in which it is not required by a masking and selective etching operation. The slice is now divided up into individual transistors and connections are made to the base and emitter regions of each transistor as shown in Fig. A.5(g). The base area of each transistor is sometimes of approximately annular shape surrounding a circular emitter area but in power transistors both base and emitter areas may be in the form of parallel strips.

The process described above produces planar transistors in large numbers and is well suited to mass production. The transistors are particularly robust and the protection of the silicon dioxide coating is such that even without sealing in cans the transistors will operate well under boiling water! Leakage currents are very low and the transistors can be designed to work up to 800 MHz. The process can also be used to manufacture junction-gate and insulated-gate field-effect transistors.

INTEGRATED (MONOLITHIC) CIRCUITS

The method of manufacture of planar transistors lends itself well to the production of integrated circuits. These are circuits designed to carry out a particular function, e.g. a bistable multivibrator or operational amplifier and may embody several transistors (bipolar or field-effect), diodes, resistors and all the necessary interconnections. All are produced on a single chip of silicon measuring perhaps less than $\frac{1}{4}$-in square by the photo-lithographic, masking, diffusion and evaporation processes described above. Resistors are areas of p-type or n-type silicon, the dimensions and impurity concentration being chosen to give the required value of resistance. Small capacitors, of the order of a few pF can consist of reverse-biased pn junctions as suggested in Chapter 1. It is inconvenient, however, to make large capacitances and integrated circuits are usually designed with direct couplings, often using emitter followers, to avoid the need for such capacitances. This is illustrated in the examples of monolithic circuits described in earlier chapters of this book.

Integrated circuits are now extensively used in linear and pulse equipment and their use is introducing a new design philosophy into electronics. Designers using monolithic devices are little concerned with components such as resistors, diodes and transistors: their main interest is in the function of the integrated circuits and their task is to choose devices with the functions necessary to carry out the purpose of the equipment and to combine them, ensuring that the impedances, signal levels, gains, bandwidths, etc., are correctly matched.

Transistor Parameters

INTRODUCTION

There are three main systems of expressing the properties of a four-terminal network. In each the network is regarded as having an

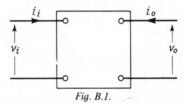

Fig. B.1.

input voltage v_i, an input current i_i, an output voltage v_o and output current i_o as illustrated in Fig. B.1.

Z PARAMETERS

In the first system the input and output voltages are expressed in terms of the input and output currents thus:

$$v_i = z_i i_i + z_r i_o \tag{B.1}$$

$$v_o = z_f i_i + z_o i_o \tag{B.2}$$

The factors z_i, z_r, etc., have the nature of impedances because, when multiplied by currents, they give voltages. The impedances can, in fact, be represented in the equivalent circuit for the transistor shown in Fig. B.2. z_i represents the input impedance of the transistor and z_o the output impedance. z_f can be called the forward impedance and z_r the reverse impedance.

At low frequencies the impedances can usually be replaced by resistances and the two fundamental equations become

$$v_i = r_i i_i + r_r i_o \tag{B.3}$$

$$v_o = r_f i_i + r_o i_o \tag{B.4}$$

in which the factors are known as r parameters.

z and r parameters are not greatly favoured largely because of

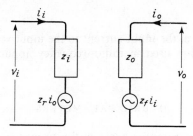

Fig. B.2.

the difficulty of measuring them. y and h parameters are generally preferred.

Y PARAMETERS

In this second system the input and output currents are expressed in terms of the input and output voltages thus:

$$i_i = y_i v_i + y_r v_o \tag{B.5}$$

$$i_o = y_f v_i + y_o v_o \tag{B.6}$$

The factors y_i, y_r, etc., have the nature of admittances because when multiplied by voltages, they give currents. The admittances can be represented in the equivalent circuit for the transistor shown in Fig. B.3.

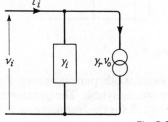

Fig. B.3.

The y parameters are more easily measured than the z parameters. If the output voltage v_o is made zero (to alternating signals) by connecting a low-reactance capacitor across the output terminals, we then have, putting $v_o = 0$ in Eqn B.5

$$i_i = y_i v_i$$

giving

$$y_i = \frac{i_i}{v_i}$$

i.e. y_i is the ratio of the input current to the input voltage.

y parameters are used in radio-frequency applications of transistors.

HYBRID PARAMETERS

These parameters combine some of the properties of the z and y parameters. In this system the input voltage and output current are both expressed in terms of input current and output voltage. The fundamental equations are as follows:

$$v_i = h_i i_i + h_r v_o \qquad (B.7)$$

$$i_o = h_f i_i + h_o v_o \qquad (B.8)$$

By inspection of these equations we can see that h_i has the dimensions of an impedance and h_o has the dimensions of an

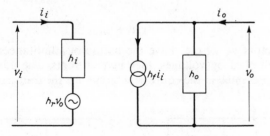

Fig. B.4.

admittance. h_r and h_f are, however, both pure numbers. For this reason these parameters are known as hybrid. The parameters can be represented in an equivalent circuit for the transistor such as that shown in Fig. B.4.

The *h* parameters are quite simple to measure. For if the output terminals are short-circuited we have $v_o = 0$ and from Eqn B.7 we have

$$h_i = \frac{v_i}{i_i} \qquad \text{(B.9)}$$

whilst from Eqn B.8 we have

$$h_f = \frac{i_o}{i_i} \qquad \text{(B.10)}$$

If the input terminals are open-circuited $i_i = 0$ and from Eqn B.7 we have

$$h_r = \frac{v_i}{v_o} \qquad \text{(B.11)}$$

whilst from Eqn B.8 we have

$$h_o = \frac{i_o}{v_o} \qquad \text{(B.12)}$$

These parameters give useful information on the transistor performance. For example from Eqn B.9 h_i is equal to the transistor input impedance for short-circuited output. From Eqn B.12 the reciprocal of h_o is equal to the output impedance for open-circuited input. From Eqn B.10 h_f gives the current gain of the transistor. From Eqn B.11 h_r is equal to the ratio of the input voltage to the output voltage: this is known as the *voltage feedback ratio* of the transistor. These parameters are often quoted in transistor manufacturers' data sheets.

It is easily possible to convert from one set of parameters to another. For example if v_o is put equal to 0 in Eqns B.5 and B.7 we can easily show that

$$h_i = \frac{1}{y_i}$$

The parameters for a particular type of circuit configuration are generally distinguished by placing a suffix *b* for common-base, *e* for common-emitter and *c* for common-collector connection although the last-mentioned is not greatly used. For example h_{ib} is the input resistance (for short-circuited output) of a common-base circuit and h_{fe} is the current gain of a common-emitter circuit.

RELATIONSHIP BETWEEN HYBRID PARAMETERS AND THE T-SECTION EQUIVALENT CIRCUIT

From Eqn 3.7, we know that the input resistance of a common-base transistor circuit, for short-circuited output terminals, is given by

$$r_i = r_e + r_b/\beta$$

This, by definition, is the parameter h_{ib}.

$$\therefore h_{ib} = r_e + r_b/\beta \qquad (B.13)$$

and, as we have seen, a typical value for this quantity is 31 Ω.

Also from Eqn 3.2 we know that the current gain for short-circuited output terminals is given by

$$\frac{i_c}{i_e} = \frac{r_b + \alpha r_c}{r_b + r_c}$$

This, by definition, is h_{fb}.

$$\therefore h_{fb} = \frac{r_b + \alpha r_c}{r_b + r_c}$$

Normally, of course, r_b is very small compared with αr_c and r_c. We can therefore write

$$h_{fb} = \alpha \qquad (B.14)$$

and a typical value for h_{fb} is 0·98.

The output resistance for open-circuited input terminals is given by Eqn 3.14. Thus we have

$$r_o = r_b + r_c$$

This is, by definition, equal to the reciprocal of h_{ob}. Thus

$$h_{ob} = \frac{1}{r_b + r_c} \qquad (B.15)$$

As r_b is very small compared with r_c this may be simplified to

$$h_{ob} = \frac{1}{r_c}$$

A typical value for r_c is 1 MΩ and thus h_{ob} is equal to 10^{-6} mho or 1 μmho.

Finally the ratio of input voltage to output voltage for open-circuited input terminals is given by

$$h_{rb} = \frac{r_b}{r_b + r_c} \qquad (B.16)$$

This result can be obtained by inspection from Fig. 2.10 on page 30. Normally r_b may be neglected in comparison with r_c and this result can thus be simplified to

$$h_{rb} = \frac{r_b}{r_c}$$

Typical values of r_b and r_c are 300 Ω and 1 MΩ, giving h_{rb} as 3×10^{-4}.

The Stability of a Transistor Tuned Amplifier

Fig. C.1 represents a transistor with input and output tuned circuits damped by parallel-connected resistances. As explained on page 168 feedback occurs via the internal collector-base capacitance c_{re} and this can cause oscillation at a frequency below the resonance value for the tuned circuits. Oscillation is most likely when the

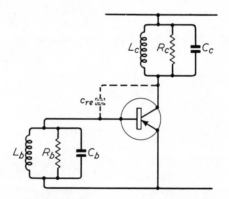

Fig. C.1. Essential features of a transistor tuned amplifier

phase shift introduced by the tuned circuits totals $90°$: if the circuits are similar and resonant at the same frequency oscillation occurs at the frequency for which each circuit gives $45°$ phase shift. From this information it is possible to arrive at a simple expression for the stability of the amplifier.

314

Fig. C.1 may be redrawn as in Fig. C.2 in which Z_f is the impedance of c_{re}. An alternating base input voltage v_b gives rise to a collector current of $g_m v_b$. This current flows through the transistor load which

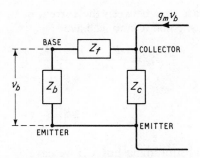

Fig. C.2. Equivalent circuit for Fig. C.1

is composed of Z_c in parallel with the series combination of Z_f and Z_b. Thus the voltage generated across Z_c is given by

$$g_m v_b \cdot \frac{Z_c(Z_f + Z_b)}{Z_c + Z_f + Z_b}$$

Z_f and Z_b constitute a potential divider across Z_c and the voltage appearing across Z_b is equal to

$$g_m v_b \cdot \frac{Z_c(Z_f + Z_b)}{Z_c + Z_f + Z_b} \cdot \frac{Z_b}{Z_f + Z_b}$$

$$= g_m v_b \cdot \frac{Z_c Z_b}{Z_c + Z_f + Z_b}$$

Now c_{re} is normally a very small capacitance and its reactance is large compared with Z_c and Z_b. Thus the voltage across Z_b is given approximately by

$$g_m v_b \cdot \frac{Z_c Z_b}{Z_f}$$

If this is equal to v_b, oscillation can occur. The condition for oscillation is thus

$$\frac{g_m Z_c Z_b}{Z_f} = 1 \tag{C.1}$$

Now Z_c is composed of R_c in parallel with X_c the net reactance of L_c and C_c. We can thus say

$$Z_c = \frac{R_c jX_c}{R_c + jX_c}$$

To give 45° phase shift between the current in Z_a and the voltage across it, X_c must equal R_c and we have

$$Z_c = \frac{jR_c}{1+j} = \frac{j(1-j)R_c}{2}$$

Similarly

$$Z_b = \frac{j(1-j)R_b}{2}$$

$Z_f = 1/j\omega c_{re}$. Substituting in Eqn C.1 we have

$$\frac{j\omega c_{re}g_m j(1-j)R_c j(1-j)R_b}{4} = 1$$

which on simplification gives

$$\omega c_{re}g_m R_b R_c = 2$$

The condition for stability is thus

$$\omega c_{re}g_m R_b R_c < 2$$

Index